Herbicides in War
The Long-term Ecological and Human Consequences

D0068699

sipri

Stockholm International Peace Research Institute

SIPRI is an independent institute for research into problems of peace and conflict, especially those of arms control and disarmament. It was established in 1966 to commemorate Sweden's 150 years of unbroken peace.

The institute is financed by the Swedish Parliament. The staff, the Governing Board and the Scientific Council are international.

Governing Board

Rolf Björnerstedt, Chairman (Sweden)
Egon Bahr (FR Germany)
Francesco Calogero (Italy)
Tim Greve (Norway)
Max Jakobson (Finland)
Karlheinz Lohs
   (German Democratic Republic)
Emma Rothschild (United Kingdom)
The Director

Director

Frank Blackaby (United Kingdom)

sipri

Stockholm International Peace Research Institute
Bergshamra, S-171 73 Solna, Sweden
Cable: Peaceresearch Stockholm
Telephone: 46/8-55 97 00

Herbicides in War
The Long-term Ecological and
Human Consequences .

Edited by
Arthur H. Westing

sipri

Stockholm International Peace Research Institute

Taylor & Francis
London and Philadelphia
1984

UK   Taylor & Francis Ltd, 4 John St, London WC1N 2ET

USA   Taylor & Francis Inc., 242 Cherry St, Philadelphia,
              PA 19106–1906

Copyright © SIPRI 1984

British Library Cataloguing in Publication Data

Herbicides in war: the long-term ecological and human consequences.
  1. Herbicides—War use.
  I. Westing, Arthur H. II. Stockholm International Peace Research Institute
  358'.38   UG447.

ISBN 0-85066-265-6

Library of Congress Cataloging in Publication Data
Main entry under title:

Herbicides in war.

  Bibliography: p. 000.
  Includes index.
  1. Herbicides—Environmental aspects—Addresses, essays, lectures. 2. Herbicides—Environmental aspects—Vietnam—Addresses, essays, lectures. 3. Herbicides—Toxicology—Addresses, essays, lectures. 4. Herbicides—Toxicology—Vietnam—Addresses, essays, lectures. 5. Herbicides—War use—Addresses, essays, lectures. 6. Herbicides—Vietnam—War use—Addresses, essays, lectures. I. Westing, Arthur H. II. Stockholm International Peace Research Institute.
QH545.P4H47 1984   574.5'2642'09597   84-2468
ISBN 0-85066-265-6

Typeset by The Lancashire Typesetting Co. Ltd,
Bolton, Lancashire BL2 1DB
Printed in Great Britain by Taylor & Francis (Printers) Ltd,
Basingstoke, Hants.

Herbicides in War: the Long-term Ecological and Human Consequences is an outgrowth of an independent 'International Symposium on Herbicides and Defoliants in War: the Long-term Effects on Man and Nature' that was held in Ho Chi Minh City, 13–20 January 1983. Neither SIPRI nor UNEP played a role in either the organization or the financial support of that Symposium. This volume has been prepared by SIPRI as a project within the SIPRI/UNEP programme on 'Military activities and the human environment'.

Preface

This book is the first product of a research programme into the environmental consequences of military activity. The programme is jointly financed by SIPRI and the United Nations Environment Programme (UNEP) and is to extend over a number of years. It is directed by Professor Arthur H. Westing of Hampshire College in Amherst, Massachusetts. SIPRI and UNEP are fortunate in that Professor Westing, a world authority on these questions, was willing to come to Stockholm to extend his studies on this subject.

Those concerned with the environment are inevitably also concerned with the effects of war, and indeed of military activity in peace-time as well. Environmentalists have to take a world view of things: so many of the factors damaging the environment—river pollution, acid rain—are no respecters of national boundaries. So they are forced, intellectually, to be world citizens, looking for international co-operation to deal with the prevention of world environmental damage. It is natural, therefore, that the preoccupation of states with massive preparations for mutual slaughter should seem to them a terrible mistake. The environmental approach to the study of military activity is a highly valuable one, simply because it is a very different way of thinking from that of the strategists concerned with geo-politics.

Although this book is produced as a SIPRI publication, with financial assistance from UNEP, much of the basic material derives from a conference organized by Professor Westing which predated his employment at SIPRI under the SIPRI/UNEP agreement. However, the subject matter fits precisely into the general framework of the study of the environmental consequences of military activity. This book is, and will remain for a long time, the definitive work on the long-term ecological and human consequences of herbicides in war.

Acknowledgements
The research assistance of Carol Stoltenberg-Hansen, the editorial assistance of Barbara Adams and the secretarial assistance of Deborah Flynn-Danielsson are gratefully acknowledged.

SIPRI   *Frank Blackaby*
March 1984   Director

Contents

Glossary and units of measure

Herbicides

Agent Blue = US Department of Defense code name for cacodylic acid (dimethyl arsinic acid; 371.5 kg/m³), altogether weighing 1 310 kg/m³; a herbicide.

Agent Orange = US Department of Defense code name for an about 1:1 mixture of 2,4,5-T (2,4,5-trichlorophenoxyacetic acid; 545.4 kg/m³) and 2,4-D (2,4-dichloro-phenoxyacetic acid; 485.1 kg/m³), altogether weighing 1 285 kg/m³; a herbicide; associated with the 2,4,5-T moiety is the impurity dioxin (2,3,7,8-tetrachloro-dibenzo-*p*-dioxin).

Agent White = US Department of Defense code name for an about 4:1 mixture of 2,4-D (2,4-dichlorophenoxyacetic acid; 239.7 kg/m³) and picloram (4-amino-3,5,6-trichloropicolinic acid; 64.7 kg/m³), altogether weighing 1 150 kg/m³; a herbicide.

2,4-D = 2,4-Dichlorophenoxyacetic acid; a herbicide; a component of Agents Orange and White.

Dioxin = 2,3,7,8-Tetrachlorodibenzo-*p*-dioxin; sometimes referred to as 'TCDD'; a strong multi-acting poison; an impurity of 2,4,5-T (2,4,5-trichlorophenoxyacetic acid), and so on.

2,4,5-T = 2,4,5-Trichlorophenoxyacetic acid; a herbicide; a component of Agent Orange; associated with 2,4,5-T is the impurity dioxin (2,3,7,8-tetrachloro-dibenzo-*p*-dioxin).

TCDD = See 'dioxin'.

Biota

For the scientific names of the biota (plants, mammals, birds, fish, insects and micro-organisms) mentioned in the text, see appendix 2.

Units of measure

The units of measure and prefixes (and the abbreviations) employed in the text are in accordance with the international system (SI) of units (Goldman & Bell, 1981). Those mentioned in the text follow:

are (a) = 100 square metres = 1 076.39 square feet.

centi- (c-) = 0.01 × .

day (d) = 24 hours = 86 400 seconds.

degree Celsius (°C) = 1 kelvin (K). To convert temperature in degrees Celsius to temperature in degrees Fahrenheit, multiply by 1.8 and then add 32.

gram (g) = 10^{-3} kilogram = $2.204\ 62 \times 10^{-3}$ pound.

hectare (ha) = 10^4 square metres = 0.01 square kilometre = 2.475 105 acres.

hect(o)- (h-) = $100 \times$.

hour (h) = 3 600 seconds.

kilo- (k-) = $10^3 \times$.

kilogram (kg) = 2.204 62 pounds.

kilometre (km) = 10^3 metres = 0.621 371 mile.

litre (L) = 10^{-3} cubic metre = 0.264 172 US gallon = 0.219 969 British gallon.

metre (m) = 3.280 84 feet.

metre, cubic (m^3) = 10^3 litres = 264.172 US gallons = 219.969 British gallons.

micro- (μ-) = $10^{-6} \times$.

milli- (m-) = $10^{-3} \times$.

minute (min) = 60 seconds.

nano- (n-) = $10^{-9} \times$.

Normal (N) = molecular weight in grams per litre of water.

part(s) per billion (ppb) = part(s) by weight per 10^9 parts.

part(s) per million (ppm) = part(s) by weight per 10^6 parts.

part(s) per trillion (ppt) = part(s) by weight per 10^{12} parts.

pico- (p-) = $10^{-12} \times$.

second (s) = see Goldman & Bell (1981, p. 3).

tonne (t) = 10^3 kilograms = 1.102 31 US (short) tons = 0.984 207 British (long) ton.

Reference

Goldman, D. T. and Bell, R. J. (eds), 1981, *International System of Units (SI)*. Washington: US National Bureau of Standards Special Publ. No. 330, 48 pp.

Introduction

Chemical weapons of one sort or another have to date been massively employed in only two wars: some 100 million kilograms of anti-personnel agents during World War I; and perhaps 91 million kilograms of anti-plant agents during the Second Indochina War. The present volume focuses on the latter of these two instances, now that more than a decade has elapsed since their employment. In doing so, it provides the only detailed exposition of the long-term effects of chemical anti-plant warfare agents based on direct on-site investigations.

This volume is the outgrowth of an independent 'International Symposium on Herbicides and Defoliants in War: the Long-term Effects on Man and Nature' that was held in Ho Chi Minh City, 13–20 January 1983 (appendix 3). That meeting brought together 72 distinguished scientists (30 ecologists and 42 physiologists) from 20 countries to interact on an informal basis with 56 of their Vietnamese counterparts (25 ecologists and 31 physiologists).[1] The primary business of the Symposium was conducted in eight working groups (4 ecological and 4 physiological), established for the purpose of discussing in depth Vietnamese researches into the long-term effects of the chemical war.

Following an initial summary chapter, the present volume is comprised of a series of eight chapters that draw upon the eight working groups of the Symposium. Each of the eight chapters opens with an evaluative summary statement which was produced by the corresponding working group during the course of the symposium. These working group summaries are followed by one or two of the most important papers on the subject which were presented at the Symposium, with strong preference here given to the Vietnamese presentations.[2] Each of the eight chapters then concludes with an overview of all the Vietnamese Symposium presentations in that particular field, made in the light of the world literature on that subject; these were prepared for this volume following the Symposium. The overview authors were in an excellent position to carry out their task not only owing to their pre-eminent status in their particular fields, but also because they had, for the most part, served as working group rapporteurs during the Symposium. Recommendations for the directions that

---

[1] In addition to the 128 participants there were 28 observers at the Symposium, including representatives of the Food and Agriculture Organization of the United Nations (FAO), United Nations Educational, Scientific and Cultural Organization (Unesco), and United Nations Environment Programme (UNEP).
[2] The nine Vietnamese papers appearing in this volume were selected from among the 29 Vietnamese papers presented at the Symposium; the two international papers (which were selected to fill gaps) were selected from among the 36 international papers presented.

future work might take in Viet Nam can be found in all of the chapters, especially in the 'Overview' papers in chapters 2 to 9. The overall summary produced at the Symposium can be found at the end of the volume (appendix 4). The volume also has appended to it a selective bibliography of pertinent background sources from the world literature (appendix 1). Additionally included are the scientific names of all biota mentioned in the text (appendix 2).

Chapter One
Herbicides in War:
Past and Present

Arthur H. Westing

Chapter 1. Herbicides in war: past and present

Arthur H. Westing
Stockholm International Peace Research Institute

I. Introduction

The Second Indochina War of 1961–1975 is noted for the widespread and severe environmental damage inflicted upon its theatre of operations, especially in the former South Viet Nam (Westing, 1976; 1980, chapter 3; 1982a). The US strategy against South Viet Nam *inter alia* involved massive rural area bombing, extensive chemical and mechanical forest destruction, large-scale chemical and mechanical crop destruction, wide-ranging chemical anti-personnel harassment and area denial, and enormous forced population displacements. In short, this US strategy represented the intentional disruption of both the natural and human ecologies of the region.

The Second Indochina War was innovative in that a great power attempted to subdue a peasant army through the profligate use of technologically advanced weapons and methods. One can readily understand that the outcome of more than a decade of such war in South Viet Nam and elsewhere in the region resulted not only in heavy direct casualties, but also in long-term medical sequelae. By any measure, however, its main effects were a widespread, long-lasting, and severe disruption of forestlands, of perennial cropland, and of farmlands—that is to say, of millions of hectares of the natural resource base essential to an agrarian society.

This chapter first reviews the history of the use of herbicides in warfare. It goes on to summarize the employment of these agents during the Second Indochina War against forest trees and crop plants and then describes their immediate effect on flora, fauna and humans. Following a brief treatment of the persistence of the agents used, it concludes with summaries of the long-term ecological and physiological effect primarily with reference to South Viet Nam, which are based largely on the chapters that follow. A select bibliography of relevant background materials is provided elsewhere in this volume (appendix 1).

II. History

Humans, in common with all animals, are dependent upon the food and shelter they derive from the plant kingdom. The intentional military destruction during war of vegetation in territory under actual or potential enemy control is a recognition of this fundamental relationship. Indeed, crop destruction has been a continuing part of warfare for millennia (Westing, 1981a), and the military importance of forests has also long been recognized (Clausewitz, 1832–1834, pages 426, 530).

Vegetational destruction can be accomplished via high explosives, fire, tractors and other means. The account below describes the employment of chemical agents for this purpose. Indeed, the sporadic use of plant-killing chemicals during both peace and war is thousands of years old. Abimelech, the son of Jerubbaal and an ancient prophet and king of Israel, is recorded in the Bible as having sowed the conquered city of Shechem (at or near the modern Nablus, Jordan; about 50 km north of Jerusalem) with salt as the final, perhaps symbolic, act in its destruction (*Judges* 9:45). The ancient Romans seem also to have employed salt in this way (Scullard, 1961, page 307).

Various inorganic herbicides (including arsenicals) have been in routine agricultural and horticultural use since the late 19th century and a number of organic ones since the mid-1930s. However, the most important advance to date in the development of herbicides was the discovery of the remarkable utility of the phenoxy and other plant-hormone-mimicking chemicals. It is thus interesting to note that their development as herbicides was tied to the then secret chemical warfare research carried out by the British and US governments during World War II (Peterson, 1967).

More than a thousand chemicals were tested in the USA during World War II in the hopes of perfecting militarily usable crop-destroying chemicals (Merck, 1946; 1947; Norman, 1946). Clearly the single most important herbicidal compound developed during this period was 2,4-D, still the most widely used herbicide in the world. Its less used and more controversial cousin, 2,4,5-T, was developed in the same way during this period. Although the possibility was considered, herbicides were not used for military purposes during World War II.

It fell to the UK in its attempt to suppress an insurgency in Malaya to be the first to employ modern herbicides for military purposes, primarily during the mid-1950s (Clutterbuck, 1966, page 160; Henderson, 1955; Henniker, 1955, page 180). The chemical anti-plant agents were used for two different purposes in this desultory decade-long war. Some of the herbicidal attacks were for defoliation along lines of communication in order to reduce the possibility of ambushes, whereas others were for the destruction of crops which were presumably being grown by or for the insurgents. These applications (both by air and from the ground) were relatively modest and rather short-lived. The major agent employed appears to have been a mixture of 2,4,5-T, 2,4-D and trichloroacetic acid (Connor & Thomas, 1984).

The only really large-scale military use of herbicides was by the USA in pursuing the Second Indochina War. This programme, the details of which are presented below, began on a very small scale in 1961, increased to a crescendo in 1969, and finally ended during 1971. Although the major US effort was directed against forests, a continuing aspect of the programme from beginning to end was crop destruction and food denial. Through the years, the US herbicide

4

spraying was confined largely to South Viet Nam, but a modest fraction of eastern Kampuchea was also involved once in 1969 (Westing, 1972). Moreover, the USA also carried out a series of herbicide missions against Laos (Westing, 1981b) and possibly a few against North Viet Nam as well (Agence France Presse, 1971).

Other than the above-noted instances, herbicides have been associated with other theatres of war and with the armed forces of other nations to only a very limited extent. For example, in 1972 the Israeli Army used 2,4-D on one occasion for crop destruction in Aqaba, Jordan (about 40 km north-north-east of Jerusalem) (Holden, 1972) and thus amazingly close to the Shechem mentioned earlier where one of the first military uses of herbicides may have occurred some 3 000 years ago.

III. The Second Indochina War

During the Second Indochina War, the USA carried out a massive herbicidal programme that stretched over a period of a decade. Although the USA was neither the first nor the only nation to employ chemical anti-plant agents as weapons of war, the magnitude of this programme was without precedent. It was aimed for the most part at the forests of South Viet Nam and, to a lesser extent, at its crops. Herbicidal attacks upon the other nations of Indochina were modest in comparison. Using a variety of agents, the USA eventually expended a volume of more than 72 million litres (91 million kilograms), containing almost 55 million kilograms of active herbicidal ingredients.

The major anti-plant agents that were employed by the USA in Indochina were colour-coded 'Orange', 'White' and 'Blue' (table 1.1). Agents Orange and White consist of mixtures of plant-hormone-mimicking compounds which kill by interfering with the normal metabolism of treated plants; Agent Blue, on the other hand, consists of a desiccating compound, which kills by preventing a plant from retaining its moisture. Agents Orange and White are particularly suitable for use against dicotyledonous plants, whereas Agent Blue is relatively more suitable for use against monocotyledonous plants. At the high levels that were used for military application—28 L/ha, averaging 21 kg/ha in terms of active ingredients (i.e., 20 to 40 times higher than normal civil usage)—these herbicides are, however, not as selective as one might expect on the basis of civil experience.

Of the several agents used, Agent Orange represented 61 per cent of the total volume expended over the years (table 1.2). The three peak years of herbicide spraying—1967 to 1969—were about equal in magnitude and together accounted for over three-quarters of the volume of total US wartime expenditures. These were also very active war years in other respects, as evidenced, for example, by the heavy US munition expenditures during this period and the high numbers of US fatalities.

Forest destruction was generally accomplished through the use of Agents Orange or White. Conversely, Agent Blue was usually the agent of choice for the

Table 1.1. Major chemical anti-plant agents employed by the USA in the Second Indochina War

Type	Composition	Physical properties	Application
Agent Orange[a]	A 1.124:1 mixture (by weight) of the *n*-butyl esters of 2,4,5-tri-chlorophenoxyacetic acid (2,4,5-T) (545.4 kg/m³ acid equivalent) and 2,4-dichlorophenoxyacetic acid (2,4-D) (485.1 kg/m³ acid equivalent); also containing 2,3,7,8-tetrachloro-*p*-doxin (dioxin) (an estimated 3.83 g/m³)	Liquid; oil soluble; water insoluble; weight 1 285 kg/m³	Applied undiluted at 28.06 L/ha, thereby supplying 15.31 kg/ha of 2,4,5-T and 13.61 kg/ha of 2,4-D, in terms of acid equivalent; and also an estimated 107 mg/ha of dioxin
Agent White	A 3.882:1 mixture (by weight) of the tri-iso-propanolamine salts of 2,4-dichlorophenoxy-acetic acid (2,4-D) (239.7 kg/m³ acid equivalent) and 4-amino-3,5,6-trichloropicolinic acid (picloram, 'Tordon') (64.7 kg/m³ acid equivalent)	Aqueous solution; oil insoluble; weight 1 150 kg/m³	Applied undiluted at 28.06 L/ha, thereby supplying 6.73 kg/ha of 2,4-D and 1.82 kg/ha of picloram in terms of acid equivalent
Agent Blue	A 2.663:1 mixture (by weight) of Na dimethyl arsenate (Na cacodylate) and dimethyl arsinic (cacodylic) acid (together 371.46 kg/m³ acid equivalent)	Aqueous solution; oil insoluble; weight 1 310 kg/m³	Applied undiluted at 28.06 L/ha, thereby supplying 10.42 kg/ha, in terms of acid equivalent (of which 5.66 kg/ha is elemental arsenic)

[a] Numerous herbicidal formulations have been tested by the USA as chemical anti-plant agents, several of which were assigned a colour code during their more or less ephemeral existence: 'Orange II' was similar to the 'Orange' above, except that its 2,4,5-T moiety was replaced by the iso-octyl ester of 2,4,5-T; 'Pink' was a mixture of the *n*-butyl and iso-butyl esters of 2,4,5-T; 'Green' consisted entirely of the *n*-butyl ester of 2,4,5-T; and 'Purple' was a mixture of the *n*-butyl ester of 2,4-D and the *n*-butyl and iso-butyl esters of 2,4,5-T.

Source: Westing (1976, page 25) adjusted in accord with Westing (1982b).

destruction of rice and other crops, although Agent Orange was much used for this purpose as well (table 1.3). All told, about 86 per cent of the missions were directed primarily against forest and other woody vegetation and the remaining 14 per cent primarily against crop plants.

Total geographic coverage of the spray missions was less than one might expect on the basis of the total expenditure of herbicides since about 34 per cent of the target areas were chemically attacked more than once during the course of the war (table 1.4). Thus the total area subjected to spraying one or more times came to an estimated 1.7 million hectares, this area being treated 1.5 times on

Table 1.2. US herbicide expenditures in the Second Indochina War: a breakdown by agent and year[a] (in m³ = 10³ L)

Year	Agent Orange[b]	Agent White[c]	Agent Blue[d]	Total[e]
1961	?	0	?	?
1962	56	0	8	65
1963	281	0	3	283
1964	948	0	118	1 066
1965	1 767	0	749	2 516
1966	6 362	2 056	1 181	9 599
1967	11 891	4 989	2 513	19 394
1968	8 850	8 483	1 931	19 264
1969	12 376	3 572	1 309	17 257
1970	1 806	697	370	2 873
1971	0	38	?	38
Total[f]	44 338	19 835	8 182	72 354

[a] To convert any of the herbicide volume data given to area covered in hectares (not considering overlap), multiply by 35.6.
[b] To convert any of the Agent Orange volume data given to total weight in kilograms, multiply by 1 285; similarly for 2,4,5-T content in kilograms, multiply by 545; similarly for 2,4-D, multiply by 485; similarly for a dioxin estimate, multiply by 0.003 83.
[c] To convert any of the Agent White volume data given to total weight in kilograms, multiply by 1 150; similarly for 2,4-D content in kilograms, multiply by 240; similarly for picloram, multiply by 64.7.
[d] To convert any of the Agent Blue volume data given to total weight in kilograms, multiply by 1 310; similarly for dimethyl arsinic (cacodylic) acid in kilograms, multiply by 371; similarly for elemental arsenic, multiply by 202.
[e] To convert any of the **Total** herbicide volume data given to average total weight in kilograms, multiply by 1 251; similarly for average kilograms of active ingredients, multiply by 757.
[f] The following amounts were sprayed in terms of active ingredients: 2,4-D, 26 million kilograms; 2,4,5-T, 24 million kilograms (containing about 170 kg dioxin); picloram, 1.3 million kilograms; dimethyl arsinic (cacodylic) acid, 3.0 million kilograms (of which elemental arsenic represents 1.7 million kilograms): total active ingredients, 55 million kilograms.
Source: Westing (1976, page 26).

Table 1.3. US herbicide expenditures in the Second Indochina War: a breakdown by type of mission and agent[a] (in m³ = 10³ L)

Type of mission	Agent Orange	Agent White	Agent Blue	Total
Forest	39 816	19 094	1 684	60 594
Miscellaneous woody vegetation	709	529	312	1 550
Crop	3 813	212	6 185	10 210
Total	44 338	19 835	8 182	72 354

[a] The same conversions provided in table 1.2, notes *a*, *b*, *c*, *d*, and *e* are also applicable to this table.
Source: Westing (1976, page 27).

Table 1.4. US herbicide expenditures in the Second Indochina War: a breakdown by number of repeat sprayings within the area covered

Number of sprayings of one area	Ultimate herbicide expenditure[a] ($m^3 = 10^3$ L)	Area involved[b] (10^3 ha)
One	31 572	1 125
Two	21 431	382
Three	11 412	136
Four	5 335	48
Five or more	2 603	19
Total	**72 354**	**1 709**

[a] To convert any of the herbicide volume data given to average total weight in kilograms, multiply by 1 251; similarly for average kilograms of active ingredients, multiply by 757.
[b] Based on the standard application rate of 28.1 L/ha. Had no area been sprayed more than once, then the total coverage would have been $2\,578 \times 10^3$ ha. As it was, the areas which were sprayed received an overall average of 42.3 L/ha, that is, they were sprayed an average of 1.51 times.

Source: Westing (1976, page 28).

the average, thereby receiving an average dose of about 42 L/ha, or about 32 kg/ha in terms of active ingredients.

Most of the anti-plant chemicals—in the neighbourhood of 95 per cent— were dispensed from C-123 (UC-123) transport aircraft equipped to deliver somewhat over 3 000 L onto 130 ha or so. The high-pressure nozzles which were used delivered droplets having an approximate median diameter of 350 μm, so that there was reasonably little drift as long as wind speeds exceeding 5 m/s were avoided, as they usually were. Normal spray time for an aircraft was just over 2 min, although the entire payload could, if the need arose, be ejected (dumped) in about 30 s, and thus onto approximately 30 ha. Of the order of 50 such dumpings occurred during the war, in which the dose level became about 120 L/ha, or about 90 kg/ha in terms of active ingredients. One aircraft (one sortie) sprayed a strip roughly 150 m wide and 8.7 km long. A mission generally consisted of 3–5 aircraft flying in lateral (side-by-side) formation. Most of the 5 per cent of the anti-plant chemicals not dispensed from fixed-wing aircraft was from helicopters, although small amounts were also dispensed from trucks and boats.

It is impossible to provide an accurate regional breakdown of herbicide expenditures for all of Indochina inasmuch as the necessary information has never been made public by the USA. It is known that about 10 per cent of South Viet Nam, the hardest hit nation, was sprayed (table 1.5). Within South Viet Nam, it was a rather large region surrounding Saigon (Ho Chi Minh City)— so-called Military Region III (figure 1.1)—that was singled out for the most intensive coverage, most intensive either on a per unit area or per capita basis. Indeed, it appears that essentially 30 per cent of the land area of Military Region III was sprayed one or more times.

Table 1.5. US herbicide expenditures in the Second Indochina War: a breakdown by region

Region	Herbicide expenditure[a] (m³ = 10³ L)	Area sprayed once or more (10³ ha)	Fraction of area sprayed (per cent)	Spraying in relation to the population (L/capita)
South Viet Nam[c]	70 720	1 670	10	4.0
Military Region I	12 022	284	10	3.9
Military Region II	14 851	351	5	4.8
Military Region III	37 482	885	29	7.7 (15.9)[b]
Military Region IV	6 365	150	4	1.0
North Viet Nam	?	?	?	?
Kampuchea	34	1	–	–
Laos	1 600	38	0.2	0.6
Total	**72 354**	**1 709**	**2**	**1.6**

[a] To convert any of the herbicide volume data given to average total weight in kilograms, multiply by 1 251; similarly for average kilograms of active ingredients, multiply by 757.
[b] The parenthetical value is based on the regional population less that of Saigon.
[c] The former Military Regions are depicted, and their areas and mid-war populations provided, in figure 1.1.

Source: Westing (1976, page 29) adjusted in accord with Westing (1972, page 186) and Westing 1981b).

Immediate effect on flora and fauna

Three vegetational categories must be singled out for special attention with respect to the wartime herbicidal attacks: dense inland forest; coastal mangrove swamp; and agriculture.

Dense inland forest

Woody vegetation covers about 10.4 million hectares, or 60 per cent, of South Viet Nam, the largest single category of which is dense inland (upland) forest. The dense inland forest, extending over about 5.8 million hectares, is composed of a complex and variable floristic conglomeration. It was also the militarily most important of South Viet Nam's land categories. To begin with, it can be estimated that about 1.4 million hectares, or 14 per cent, of the total extent of South Viet Nam's woody vegetation was sprayed one or more times (table 1.6). Of this, perhaps 1.1 million hectares occurred in the dense inland forest type, which represents about 19 per cent of that vegetational category. The dense forest lands within so-called War Zones C and D (figure 1.1) were particularly hard hit.

Following herbicidal attack of an inland forest, fairly complete leaf abscission (as well as flower and fruit abscission) occurred within two or three weeks. The surviving trees usually remained bare until the onset of the next rainy season, often, therefore, for a period of several or more months. To achieve total defoliation of the lower storeys of a multiple canopy forest necessitated one or more follow-up sprayings.

Figure 1.1. South Viet Nam during the Second Indochina War
(Populations shown are estimates for 1969.)

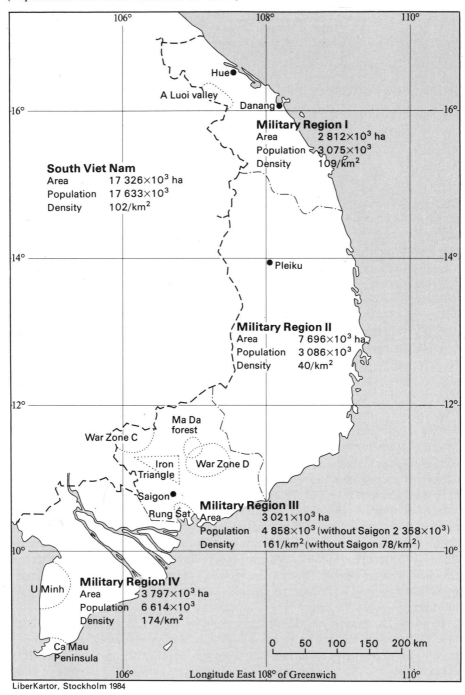

Military Region I
Area   2 812×10³ ha
Population 3 075×10³
Density   109/km²

South Viet Nam
Area   17 326×10³ ha
Population 17 633×10³
Density   102/km²

Military Region II
Area   7 696×10³ ha
Population 3 086×10³
Density   40/km²

Military Region III
Area   3 021×10³ ha
Population 4 858×10³ (without Saigon 2 358×10³)
Density   161/km² (without Saigon 78/km²)

Military Region IV
Area   3 797×10³ ha
Population 6 614×10³
Density   174/km²

Hue
A Luoi valley
Danang
Pleiku
Ma Da forest
War Zone C
Iron Triangle
War Zone D
Saigon
Rung Sat
U Minh
Ca Mau Peninsula

0   50   100 150 200 km

Longitude East 108° of Greenwich

LiberKartor, Stockholm 1984

Source: Westing (1976, pages 3, 7).

Table 1.6. US herbicide expenditures, in 10^3 ha, in South Viet Nam in the Second Indochina War: a breakdown by vegetational type

Vegetational type	Area	Area sprayed once or more
Dense forest	5 800	1 077
Primary	4 500	836
Primary plus secondary	600	111
Secondary	700	130
Open (clear) forest	2 000	100
Bamboo brake	800	40
Mangrove forest (swamp)	500	151
True	300	124
Rear (back)	200	27
Rubber plantation	100	30
Pine forest	100	0
Miscellaneous woody vegetation	1 100	36
Woody sub-total	*10 400*	*1 434*
Paddy (wet) rice	2 500	59
Field crops (upland rice, etc.)	500	177
Agricultural sub-total	*3 000*	*236*
Miscellaneous	*3 926*	*0*
Total	**17 326**	**1 670**

Source: Westing (1976, page 30) adjusted in accord with table 1.5.

Virtually all of the many dicotyledonous tree species subjected to spraying were defoliated at the intensity of treatment employed. Then, at the time of refoliation, a spectrum of sensitivity became evident among the many hundreds of tree species which comprise this vegetational type. Only about 10 per cent (some observers have suggested more) of the trees were killed outright by a single military spraying, a situation true for perhaps 66 per cent of the total sprayed area (table 1.4). The survivors displayed various levels of injury, as evidenced by differing severities of crown (branch) dieback, temporary sterility and other symptoms. Such injury in time resulted in some further delayed mortality among the survivors. Moreover, once the understory was deprived of the protection of a destroyed (or largely destroyed) overstory, some fraction of the understory trees in time dies of environmental conditions too harsh for their existence. When the military situation led to more than one herbicidal attack, as occurred on about 34 per cent of all sprayed lands, the level of tree mortality increased more or less exponentially with each subsequent spraying, more steeply so with briefer intervals between sprayings. It might also be noted that in economic terms the spraying of South Viet Nam's dense inland forest resulted in a loss of commercial timber estimated to be of the order of 20 million cubic metres.

When the trees were sprayed, causing the leaves to fall and decompose, the soil was for the most part unable to hold the released nutrients so that these were lost to the local ecosystem, a phenomenon referred to as nutrient dumping, especially acute in the tropics. Further nutrient losses and other site debilitation occurred since the death of the vegetation led to accelerated soil erosion. Both erosion and nutrient dumping continued until the area in question was again stabilized by the establishment of a replacement (pioneer) vegetation, usually during the subsequent growing season.

The principal impact on the wildlife of sprayed sites was via a diminution in the food and cover (shelter) afforded by the vegetation. Here it must be noted that a significant majority of the animal life in a tropical forest is found in—and depends upon—the upper vegetational stories, precisely the portion of the eco-·system most seriously impaired by massive herbicidal attack. There was also a more or less modest level of damage to wildlife via the direct toxic action of the herbicides sprayed. Some birds appear to have succumbed in this fashion and probably several kinds of invertebrates as well, including some aquatic inverte-brates and some terrestrial insects.

Thus it can be seen that the immediate ecological impact of military spraying of an inland forest site with anti-plant chemicals can be severe, especially if repeated. The primary producers (green plants) of an ecosystem are knocked back drastically, to be replaced by a new community of significantly lesser biomass, smaller nutrient-holding capacity, and reduced primary productivity. A poorer soil results from the attack, with a lesser fraction of humus (organic matter) and often exhibiting a chronic shortage of nitrogen. Fire subsequent to herbicidal attack would aggravate the situation. Particularly in those inland areas that were sprayed several times—some 200 thousand hectares (table 1.4)—the overstory destruction was sufficient to permit the release of, or invasion by, certain tenacious pioneer grasses. These included both herbaceous types such as *Imperata cylindrica* and woody types such as frutescent (shrubby) bamboos.

Mangrove habitat

The mangrove habitat, scattered primarily along the southerly coastline of South Viet Nam, occupies approximately 500 thousand hectares of inhospitable, seemingly impenetrable, and outwardly unimportant swamps. It is singled out here owing to the widespread and peculiarly drastic herbicidal damage it suffered during the Second Indochina War. An estimated 124 thousand hectares of true mangrove—41 per cent of that entire sub-type—plus another 27 thousand hectares of rear (back) mangrove (13 per cent of that sub-type) were subjected to military herbicide spraying during the war. Unlike the relatively modest degree of kill resulting from such action in inland forest types, even a single spraying in this coastal lowland type most often destroyed essentially the entire plant community (the herbicidal sensitivity probably being related to its osmotic versatility, i.e., to its wide tolerance to changes in salinity). Virtually nothing remained alive even after a single herbicidal attack and the resulting scene was weird and desolate. It subsequently became even worse when exacer-bated by the usual salvåge harvesting of the killed trees and by the inevitable erosion.

The taxonomically divers plant species that make up the mangrove community all displayed great sensitivity to the hormone-mimicking herbicides, with *Rhizophora*—the economically most important genus of trees—being especially sensitive. Little if any immediate recolonization occurred on the herbicide-obliterated sites. With the primary producers essentially wiped out, their energy-capturing function was lost and all else that built on this. Both food and cover were eliminated by the US attacks, affecting not only the enemy forces but the indigenous forest creatures as well. For example, Orians & Pfeiffer (1970) reported an enormous reduction in numbers of birds.

Less obviously, the reticulation of channels throughout a mangrove swamp—roughly one-quarter of the surface area of the mangrove type in South Viet Nam—supports a rich variety of aquatic fauna during all or part of their life cycle. These organisms depend directly or indirectly on a steady and enormous supply of nutrients dropped, flushed and leached out of the terrestrial part of the system. Numerous species of fish and crustaceans that spend their adult lives offshore, and some that migrate and live up the rivers, utilize the mangrove estuaries as breeding and/or nursery grounds. There were early indications of post-war declines in South Viet Nam's offshore fishery, involving both true fish and shellfish (crustaceans), attributed to the wartime spraying.

It is important to note that the mangrove type is a transitional zone between land and sea and thus appears to serve the added important function of stabilizing the shoreline. As the coastline accretes, mangroves invade the virgin lands and their roots hold the soil against the action of wind, wave, current and tide. The unprotected channel banks and barren mud flats created by the spraying were seen to erode at rapid rates.

With so much of the mangrove habitat destroyed, it was clearly the ecosystem most seriously affected by the Second Indochina War. With the promise of long-term conversion of a significant fraction of this habitat to other vegetational types (whether by natural or anthropogenic means) the question arises to what extent such long-term habitat loss will lead to species extinctions. It is known that the number of species within any particular taxonomic group that an isolated habitat can support is related to its area. If a habitat is reduced in size, as was the case with South Viet Nam's mangroves, the resulting excess of species will in due course die out. It has been estimated, for example, that a 10 per cent reduction in this mangrove habitat (a likely situation) will in time lead to a 3 or 4 per cent loss in the indigenous plant and animal species.

Agriculture

During the Second Indochina War the USA carried out a routine military policy of systematic large-scale crop destruction in South Viet Nam. Chemical crop destruction from the air made up the greatest fraction of this major resource denial programme. Chemical crop destruction is estimated to have affected some 236 thousand hectares of agricultural lands in South Viet Nam one or more times (about 8 per cent of the total) (table 1.6); in all, herbicides were sprayed over some 356 thousand hectares in South Viet Nam, although this larger value does not take into account sprayings of the same fields during different years. (At least 8 thousand hectares of crop lands were additionally sprayed elsewhere in Indochina, especially Laos.) The crop spraying is estimated to have resulted in the immediate destruction of more than 300 million kilograms of food. Additionally, perhaps 30 per cent of South Viet Nam's 135 thousand hectares of rubber plantations were destroyed by herbicides during the war.

Immediate impact on humans

The anti-plant agents employed by the USA in Indochina were selected for their herbicidal qualities. At least to begin with, none was generally known to be particularly toxic to humans. Nonetheless, during the war there were occasional seemingly authentic medical problems experienced by farmers, woodsmen and other local inhabitants subjected to military herbicide spraying (Betts & Denton, 1967, page 25; Burchett, 1963, pages 61–62; Carrier, 1974; Hickey, 1974, pages 11–15; Murphy et al., 1974, chapter 7; Orians & Pfeiffer, 1970; Rose & Rose, 1972; Tung et al., 1971; Tung, 1973; Westing, 1972, pages 196–198). Those exposed often described a series of more or less temporary ailments that included dizziness, headaches, vomiting, diarrhoea, lacrimation, coughing and dyspnoea (laboured respiration). Infants were said to suffer considerably more than adults. Some deaths were also locally attributed to the herbicides (Carrier, 1974).

It must be recalled that the routine military herbicide applications were at least an order of magnitude heavier than comparable civil ones. Moreover, owing to a number of military exigencies, the entire payload of an aircraft sometimes had to be rapidly ejected, thereby exposing a limited area to an exceedingly high dose. Such emergency dumpings occurred, as noted earlier, some 50 times during the war (about monthly during the late 1960s).

Not only was some fraction of the Vietnamese population thus exposed to relatively massive doses of herbicides, but the effect of such exposure could well have been magnified by the generally sub-optimal levels of nutrition and health of the affected people. Many if not most of those involved were malnourished, parasite ridden and otherwise ill. Some of the reported problems may have stemmed from particular sensitivities or allergies to the herbicides used. For example, it is not uncommon that exposure to dimethyl arsinic acid (the active component of Agent Blue, which accounted for 11 per cent of all spray operations) results in headaches, abdominal cramps, weakness and nausea (Tarrant & Allard, 1972). However, the most likely cause of the reported medical problems related to herbicides was exposure to Agent Orange and, more specifically, to its poisonous dioxin contaminant.

Agent Orange accounted for 61 per cent of the wartime herbicide spray operations. Its dioxin content averaged an estimated 4 g/m^3, although occasional lots contained between 10 and 20 times this amount. Thus, whereas the average delivery rate of dioxin was almost 110 mg/ha, it was occasionally 1–2 g/ha. Moreover, there can be estimated to have been of the order of 30

Agent Orange dumpings during the war, these delivering concentrations of more than four times the values just given. The question of whether this applied dioxin had a long-term health impact on the indigenous human population is considered in a subsequent section.

Human health problems that could be ascribed in a secondary sense to the military employment of herbicides should also be mentioned. Thus the herbicide operations resulted in substantial human displacement (i.e., in the generation of refugees) and thereby contributed to the spread of infectious diseases. The reduced food intake brought about by herbicidal crop destruction, both intended and incidental, resulted in both acute and chronic health problems. The war, of course, also took a heavy physiological toll among the South Vietnamese, and the herbicidal attacks contributed significantly to this burden (Murphy *et al.*, 1974, chapter 8). Additionally, the ecological disruption brought about by the herbicide operations exacerbated a number of disease problems via improved habitat conditions for various vector organisms such as *Anopheles* mosquitoes and rats.

Finally it must be noted that some of the self-reported herbicide-related health problems—although real enough—may, in fact, have had nothing to do with herbicides. First of all, some of the reports may have stemmed from coincidental disease problems first noticed during inspection elicited by herbicidal attack. Second, some of the reports can be attributed to confusion with the effects of other chemicals also liberally dispensed from the air by the USA during the war. These included the harassing anti-personnel agent CS and the insecticide malathion (table 1.7). CS, of course, brings about considerable more or less

Table 1.7. Major chemical agents sprayed from the air in the Second Indochina War: approximate gross areal coverage[a]

Agent[b]	Spray period (approximate)	Amount sprayed (10^6 kg)	Area sprayed[c] (10^6 ha)	Area sprayed[c] (per cent)
Orange	1962–1970	57.0	1.6	12
White	1966–1971	22.8	0.7	5
Blue	1962–1970	10.7	0.3	2
CS	1964–1970	9.0	5.0	37
Malathion	1967–1972	3.0	6.0	44
Total	**1962–1972**	**102.5**	**13.6**	**100**

[a] Primarily in South Viet Nam.
[b] The anti-plant (herbicidal) Agents Orange, White, and Blue are described in table 1.1. The harassing (irritant) anti-personnel Agent CS is *o*-chlorobenzalmalonitrile. The insecticide malathion is S-(1,2-dicarbethoxyethyl)-0, 0-dimethyldithiophosphate. Agents Orange, White, Blue and malathion were dispensed primarily from fixed-wing aircraft (C-123s) and the CS primarily from helicopters.
[c] Not considering multiple sprayings of the same area.

Source: The data for Agents Orange, White, and Blue are from table 1.2. Those for CS are from Westing (1976, pages 53–54). Those for malathion are from the US Department of Defense (private communication, 14 Dec 1970) and Buckingham (1982, pages 124–125).

transient malaise. Malathion can also cause transient headaches, nausea and weakness (Gardner & Iverson, 1968). A victim of wartime spraying in South Viet Nam had roughly only a one in ten chance of having been sprayed by Agent Orange and a one in five chance of having been sprayed by a herbicide.

Herbicidal persistence

Important in any consideration of the impact of herbicides is their persistence and mobility, that is, how long they will remain present and active in the soil and biota and whether they will move up in food chains, perhaps even concentrating in the process (so-called ecological amplification). For 2,4-D, representing 48 per cent of the chemicals sprayed (table 1.2), a level of environmental insignificance (as determined by lack of obvious effect on all but the most highly sensitive of subsequently planted test species) is reached within a month or so. For 2,4,5-T, representing 44 per cent of the chemicals sprayed, this occurs within five months or so; for picloram, representing 2 per cent of the chemicals sprayed, within perhaps 18 months; and for dimethyl arsinic acid, representing 6 per cent of the chemicals sprayed, after about a week.

The dioxin contaminant of the 2,4,5-T in Agent Orange, on the other hand, turned out to be considerably more persistent than its carrier agent. A conservatively estimated total of 170 kg was applied to South Viet Nam, primarily during 1966 to 1969 and largely in the former Military Region III (table 1.8). The dioxin, once incorporated into the local ecosystem, can be assumed to disappear from the environment following first-order kinetics and can be calculated to have an environmental half-life of the order of 3.5 years[1]. If one makes the simplifying assumptions that the estimated 170 kg of applied dioxin had all been introduced into the South Vietnamese environment in 1968 and that half of it had become incorporated into the soil and biota (the other half presumably having been rather rapidly photo-decomposed), then perhaps 8 kg remained present in 1980, 3 kg will be present in 1985, and 1 kg in 1990. The action of wind and water is likely to have enlarged (and continues to be enlarging) the original area of application of 1.0 million hectares, a matter that is disturbing in the sense that the area of contamination is expanding, but reassuring in the sense that the severity of contamination in any one locality is declining not only via decomposition, but also through scattering.

[1] The environmental half-life for dioxin of 3.5 years given in the text was calculated by the author from data available from a spray test site in Florida, USA (Westing 1982a, page 368). A rough estimate by Olie (paper 9.B), based on data from Viet Nam, comes to 5 years. Moreover, Domenico et al. (1980) collected 176 more or less comparable dioxin-contaminated soil samples (from the top 7 cm) during the 18 months following an industrial accident that occurred at Sèveso, Italy in 1976. These data permit one to calculate a half-life of 1.9 years, albeit an unreliable one.

Table 1.8. Dioxin applications in South Viet Nam in the Second Indochina War: a rough approximation[a]

Year	Military Region I[b]	Military Region II[b]	Military Region III[b]	Military Region IV[b]	Total
A. Amount (kg)					
1961	?	?	?	?	?
1962	–	–	0.1	–	0.2
1963	0.2	0.2	0.6	0.1	1.1
1964	0.6	0.8	1.9	0.3	3.6
1965	1.2	1.4	3.6	0.6	6.8
1966	4.1	5.1	12.9	2.2	24.4
1967	7.7	9.6	24.1	4.1	45.5
1968	5.8	7.1	18.0	3.1	33.9
1969	8.1	10.0	25.1	4.3	47.4
1970	1.2	1.5	3.7	0.6	6.9
Total	**28.9**	**35.7**	**90.1**	**15.3**	**170.0**
B. Amount per unit area, assuming uniform distribution over the entire region (mg/ha)					
1961	?	?	?	?	?
1962	–	–	–	–	–
1963	0.1	–	0.2	–	0.1
1964	0.2	0.1	0.6	0.1	0.2
1965	0.4	0.2	1.2	0.2	0.4
1966	1.5	0.7	4.3	0.6	1.4
1967	2.8	1.2	8.0	1.1	2.6
1968	2.0	0.9	5.9	0.8	2.0
1969	2.9	1.3	8.3	1.1	2.7
1970	0.4	0.2	1.2	0.2	0.4
Total	**10.3**	**4.6**	**29.8**	**4.0**	**9.8**

[a] The estimated 170 kg of dioxin was directly applied to about 1.0 million hectares, that is, onto about 6 per cent of the surface of South Viet Nam (table 1.2). Thus the average dose on this directly sprayed land was about 163 mg/ha. About 155 kg, or 91 per cent, of the applied dioxin was sprayed onto forest lands and the remaining 15 kg, or 9 per cent, onto agricultural lands (table 1.3).

[b] The former Military Regions are depicted, and their areas and mid-war populations provided, in figure 1.1.

Source: Calculated from the data in tables 1.2 and 1.5.

IV. South Viet Nam today

With more than a decade having elapsed since the herbicidal assault on Indochina, an international group of ecological and physiological scientists convened recently in southern Viet Nam in order to evaluate the available evidence of long-term effects of that event (appendix 3). The chapters that follow present the Working Group summaries plus some of the most important individual papers which emerged from that symposium. The major findings of this unique effort are summarized below, first with respect to lasting effects on nature and then to those on humans. (For a recent overview of the general environmental aftermath of the Second Indochina War, see Westing [1982a].)

Long-term effect on flora and fauna

Recent examination of the inland forests of South Viet Nam has established that the wartime herbicidal damage of more than a decade ago is still much in evidence. It was re-affirmed that the severity of original damage and progress towards recovery depend on a variety of complex (and often little understood) factors, including: pre-spray condition of the stand; frequency and season of original spraying; species composition; steepness and other features of the terrain; local climate; areal extent of damage; availability of a seed source, and subsequent fire history (see below). It appears that with one or two original sprayings of a dense inland forest a sufficient number of understory trees survived that will grow and provide at least a poor harvest in three to four decades following attack (Ashton, paper 2.C). However, it was estimated to take eight to ten decades following such spraying for a stand comparable to the pre-spray one to become established.

Those inland forests sprayed three or more times were generally damaged sufficiently to result in subsequent site damage from soil erosion and nutrient dumping (loss of nutrients in solution) and for the establishment of a grassy cover, usually herbaceous though sometimes woody (Ashton, paper 2.C; Hiêp, paper 2.B). It turns out that these now herbaceous-grass-dominated sites have been burned over during many of the annual dry seasons since the war, such fires often being of human origin (Trùng et al., paper 2.A). These repeated fires have not only essentially prevented the re-establishment of trees, but have even been encroaching on the surrounding forest and have thus been slowly expanding in size (Galson & Richards, paper 2.D). The modest natural forest regeneration in these badly damaged areas has been with poor quality trees. The very important post-war role of fire in impeding forest recovery and even in exacerbating the original degree of damage has been the major revelation of the recent inland forest studies.

Herbicidal decimation of a forest leads to site debilitation for a number of reasons. The nutrients released by the fallen foliage cannot be held to any great extent by the soil and are thus lost to that ecosystem. Such nutrient dumping is especially severe in the tropics and often prominently involves potassium, nitrogen and phosphorus (Huây et al., paper 4.A; Zinke, paper 4.C). As the trees die, the newly unprotected soil is subject to erosional loss—the more so the steeper the terrain—until the re-establishment of a new vegetational cover (a grass cover which, however, protects the soil less well than the former trees). Indeed, recent soil studies have revealed that soils on steep slopes that had been subjected to the wartime spraying are, more than a decade later, still seriously depleted in nitrogen as well as in total organic matter content (Huây & Cu, paper 4.B).

It has become quite clear that for vegetational recovery to occur in the seriously damaged inland forests fire must be excluded and, moreover, that the worst damaged areas will require artificial planting. Indeed, site debilitation has in many instances been sufficiently severe to require pre-planting (or interplanting) with hardy soil-holding and soil-enriching species, for example, with nitrogen-fixing leguminous trees (Ashton, paper 2.C; Galston & Richards, paper 2.D).

The close association between an animal's geographic distribution (i.e., the animal's presence or absence) and its particular habitat requirements is a fundamental tenet of ecology. Indeed, this relationship is an especially tight one in tropical forests (Leighton, paper 3.C). Recent comparisons in South Viet Nam of unsprayed inland forest sites with comparable sites that had been multiply sprayed during the war, have been subjected to subsequent fires, and are now dominated by grasses abundantly confirm this relationship. For example, in two unsprayed forests 145 and 170 birds were recorded whereas in the destroyed forest (now grassland) there were only 24 (Qúy et al., paper 3.A). Similar values for mammals were 30 and 55 in the two unsprayed sites, but only 5 in the comparable though previously sprayed site. Moreover, an examination of the mammalian species that comprise these numbers reveals that whereas most taxa of wildlife declined, the numbers of undesirable rodent species increased.

To ameliorate the disastrous long-term impact of destroyed habitat on wildlife populations will require an accelerated programme of reforestation, the prohibition of game hunting, and restrictions on fuel-wood gathering (Huỳnh et al., paper 3.B). More sophisticated actions are called for as well (Leighton, paper 3.C).

As noted earlier, the one habitat of South Viet Nam which had been most seriously disrupted by the wartime herbicidal attacks was mangrove. Roughly 124 000 ha of this highly productive ecosystem (i.e., about 40 per cent of it) had been utterly devastated. A rough field survey carried out in 1980 indicated the following current situation regarding these 124 000 ha (Westing, 1982a, pages 373–375): (a) barren patches 5–50 ha in size, about 5–10 per cent; (b) natural regeneration of *Rhizophora* adjacent to residual stands (a highly desirable outcome), about 1 per cent; (c) artificial planting of *Rhizophora*, about 10 per cent; (d) conversion to rice and other crops, about 5–6 per cent; and (e) natural regeneration of low-growing locally undesirable species of palms, ferns, poor-quality mangrove species, etc., about 75 per cent. Much site damage by sheet erosion and wave action has occurred since the war (Snedaker, paper 5.D). The lack of natural regeneration by the ecologically and economically desirable *Rhizophora* has to a considerable extent resulted from a lack of seed source so that an accelerated programme of artificial planting is indicated (Hiêp, paper 5.B; Snedaker, paper 5.D; Yên et al., paper 5.A). Where mangrove species have become established (whether by natural or artificial means and whether of inferior or superior species) a closed canopy can be expected within a decade or two of the time of establishment and a harvestable crop of wood (small timbers and firewood) in perhaps four or five decades.

The offshore marine fishery of South Viet Nam is known to have declined since the war (Yên et al., paper 5.A), but whether this phenomenon finds its roots in the wartime herbicidal attacks, as is suggested from time to time, has not been demonstrated (Snedaker, paper 5.D). However, one recent study indicates that freshwater fish in inland forest areas that had been attacked with herbicides during the war became and have remained substantially reduced both in species numbers and biomass (Yên & Quýnh, paper 5.C). The reduction was attributed to a long-lasting decline within the affected waters of the algae and invertebrates that provide the food for these fish.

Long-term effect on humans

It is quite difficult to establish the extent to which wartime exposure to herbicides has resulted in long-term medical problems in Viet Nam. One of the fundamental problems in attempting to clarify this matter has been to establish to an adequate degree of satisfaction the level to which—and even whether—an individual had been exposed to the herbicides (see comments above under 'Immediate effect on humans'). Separating the long-term effects of herbicides from those of any confounding concomitant health-compromising environmental factors has provided the basis for a second intractable factor (e.g., the effects of exposure to certain parasites, to hepatitis-B virus, or to aflatoxin). Grossly inadequate demographic and public health statistics and very poor hospital records have also exacerbated the problem. Insufficient numbers of medical, scientific and support personnel, not enough supplies and funding, and unsophisticated laboratory, computing, and other research facilities have combined to compromise the matter further. It is thus to the enormous credit of the Vietnamese medical and scientific establishments that much highly useful information has emerged from their recent efforts to elucidate the question, as demonstrated by the chapters that follow.

To begin with, it is clear that huge amounts of herbicides were broadcast widely in South Viet Nam during the Second Indochina War, especially so in the former Military Region III during the three years 1967–69 (tables 1.2 and 1.5). Second, it is known that Agent Orange, the most widely used of the herbicidal agents and accounting for some 60 per cent of the volume and area sprayed, contained a significant level of the highly toxic contaminant dioxin (table 1.8). And third, it has been established without question that in its introduction and incorporation into the South Vietnamese environment, dioxin has been inexorably inflicted upon large numbers of people. Human exposure to dioxin occurred both directly at the time of spraying and indirectly through the diet, owing to its lengthy environmental persistence (dioxin having an environmental half life of at least three years, and thus being to some extent still present) and to its more or less ready environmental mobility (with movement occurring up food chains which culminate in humans) (Olie, paper 9.B; Rappe, paper 9.C; Thu *et al.*, paper 9.A).

It is also thoroughly established that dioxin is an extraordinarily potent poison, producing a wide range of organ and metabolic dysfunctions at the ng/kg to μg/kg levels, in terms of mammalian body weight (Dwyer & Epstein, paper 6.D; Hay, paper 8.C; Trung *et al.*, paper 8.A). Dioxin has been shown, at least in laboratory mammals, to be carcinogenic (causing cancer) and teratogenic (causing birth defects); and laboratory tests show that it is potentially mutagenic (causing genetic, i.e., DNA or gene, damage) as well (Hay, paper 8.C).

Recent studies in South Viet Nam have established to a certain degree of satisfaction that wartime exposure to herbicides has led to a higher than expected level of a variety of persistent clinical problems that include chronic asthenia (weakness), recurring headaches, periods of depression and anxiety, gastro-intestinal disorders, nausea, and lack of libido—many of which come under the heading of neuro-intoxications (Dwyer & Epstein, paper 6.D; Trinh, paper 6.B; Truong *et al.*, paper 6.A). These effects are not too surprising in the light of the herbicide toxicology literature (Dwyer & Epstein, paper 6.D), but could also be in part attributable to other factors, including other chemicals (as alluded to earlier in this chapter). Additionally there is the suggestion that exposure may in South Viet Nam have led to a higher than normal incidence of chronic hepatitis (inflammation of the liver) (Trinh, paper 6.B). This is of special interest inasmuch as there seems to be no precedent for such an effect among herbicide-exposed humans in other parts of the world.

The possibility that phenoxy herbicides (such as those that comprise Agent Orange) and/or their dioxin contaminant give rise to a higher incidence of cancers among exposed humans is suggested by firm evidence from studies on laboratory mammals and by an apparent increase in soft-tissue sarcomas[2], in connection with occupational exposure in Sweden (Dwyer & Epstein, paper 6.D). Although there is no evidence from Viet Nam for an increase in soft-tissue cancers, both an earlier observation (Tung, 1973) and a limited recent survey (Vân, paper 6.C) point to the possibility of an increase in human liver cancers (primary hepatic carcinomas). As with the possible increase in human chronic hepatitis referred to just above, there is no corroboration outside of Viet Nam for this putative increase in human primary hepatic carcinomas. This matter warrants vigorous follow-up studies both in Viet Nam and elsewhere, especially since other factors in the environment may be confounding the issue (Moertel, 1982).

The possibility that exposure to the military herbicides would result in unfavourable reproductive effects in humans has been a major fear since the latter part of the war (Meselson *et al.*, 1972). Of particular concern here are such adverse outcomes of pregnancy as spontaneous abortions (miscarriages), still-births, molar pregnancies (in which a hydatidiform mole developes in lieu of a foetus), and congenitally malformed offspring (infants with birth defects) (Westing, paper 7.C). As noted earlier, the dioxin contaminant of Agent Orange is a known mammalian teratogen (producer of birth defects).

The evidence from Viet Nam is to date inconclusive with respect to adverse outcomes of pregnancy, although it is highly suggestive in a number of ways (Cân *et al.*, paper 7.A). For example, a summary of hospital records in South Viet Nam by Huong & Phuong (as summarized in Westing, paper 7.C) suggests

---

[2] 'Soft-tissue sarcoma' is a fleshy malignant tumour of the soft somatic tissues (Morton & Eilber, 1982). The soft somatic tissues, which make up more than 50 per cent of the body weight, include the fibrous and adipose connective tissues, blood vessels, nerves, smooth and striated muscles, fascia (ligaments, tendons and so on) and their synovial structures (sheaths and so on), and lymphatic structures. However, cancers of the lymphatic system—Hodgkin's disease (Rosenberg, 1982), Burkitt's lymphoma (Ziegler, 1982), and other malignant lymphonas (DeVita *et al.*, 1982)—are not usually subsumed under the term 'soft-tissue sarcoma'.

the possibility of a war-related increase in spontaneous abortions and perhaps as well as in molar pregnancies. An apparent (and unexpectedly persistent) increase in chromosomal aberrations in the peripheral blood of individuals presumably exposed to herbicides during the war (Trung & Dieu, paper 8.B) lends support to the possibility of an increased frequency of adverse outcomes of pregnancy owing to the known correlation between these two phenomena.

The most provocative reproductive effect to emerge from the recent Vietnamese studies is the possibility that wartime exposure to herbicides of males who then during the war mated with unexposed females led to an increase in adverse outcomes of pregnancy. It appears from one study that in such cases the frequency of spontaneous abortions went up; and that among the live births the frequency of congenital malformations went up, especially cleft lips (Cân, paper 7.B). Such a paternally transmitted effect on human reproduction appears to be unique and therefore demands corroboration.

V. Conclusion

Faced during the Second Indochina War with a dispersed and elusive enemy in South Viet Nam, the USA sought to deny this foe sanctuary, freedom of movement, and a local civilian economy from which to help derive its sustenance. This strategy was pursued, *inter alia*, through an unprecedentedly massive and sustained expenditure of herbicidal chemical warfare agents against the fields and forests of South Viet Nam. The use of these agents resulted in large-scale devastation of crops, in widespread immediate damage to the inland and coastal forest ecosystems, and in a variety of health problems among exposed humans.

The damage to nature involved the death of millions of trees and often their ultimate replacement by grasses, in turn maintained to this day by subsequent periodic fires; deep, lasting inroads into the mangrove habitat; widespread site debilitation via soil erosion and loss of nutrients in solution; decimation of terrestrial wildlife primarily via destruction of their habitat; losses in freshwater fish, largely because of reduced availability of food species; and a possible contribution to declines in the offshore fishery. The impact on the human population has included long-lasting neuro-intoxications as well as the possibility of increased incidences of hepatitis, liver cancer, chromosomal damage, and adverse outcomes of pregnancy from exposed fathers (especially spontaneous abortions and congenital malformations).

A vigorous and sustained research effort is warranted in Viet Nam in order to pursue and ameliorate the long-term ecological and medical effects of the wartime use of the herbicides. The proposed ecological studies should pursue techniques of fire prevention, soil restoration, tree planting (including pre-planting and inter-planting) and wildlife restoration. Study areas and field stations should be established in both inland and coastal habitats. The proposed medical studies should pursue the unexpected suggestions of increases in hepatitis, liver cancer, chromosomal aberrations and paternal influence on adverse outcomes of pregnancy. Both the ecological and medical studies might best be carried out with the active co-operation of such international agencies as the United Nations Educational, Scientific and Cultural Organization (e.g., with its Regional Coastal and Marine Programme) and the World Health Organization (e.g., with its International Agency for Research on Cancer).

Finally the question arises regarding future employment of herbicides as anti-plant chemical warfare agents and of the potentially ecocidal[3] outcome of their use (Westing, 1980, pages 189–191). Military evaluations have been favourable as regards a diversity of potential operational theatres (Engineers, 1972). On the other hand, a widely held interpretation of the Geneva Protocol of 1925 makes illegal their use in war. Moreover, their impact—especially as demonstrated by the Second Indochina War—makes it illegal to use them in the light of the Environmental Modification Convention of 1977.

References

Agence France Presse. 1971. Hanoi issues list of raids: defoliation flights alleged. *New York Times*, 1971 (21 Jan): 13.

Betts, R. and Denton, F. 1967. *Evaluation of Chemical Crop Destruction in Vietnam*. Santa Monica, Cal.: Rand Corp. Memo. No. RM-5446-ISA/ARPA, 13 + 34 pp.

Buckingham, W. A., Jr. 1982. *Operation Ranch Hand: the Air Force and Herbicides in Southeast Asia 1961–1971*. Washington: US Air Force Office of Air Force History, 253 pp.

Burchett, W. G. 1963. *Furtive War: the United States in Vietnam and Laos*. New York: International Publishers, 224 pp.

Carrier, J. M. 1974. *Effects of Herbicides in South Vietnam. B[4]. Estimating the Highlander Population Affected by Herbicides*. Washington: National Academy of Sciences, 13 pp.

Clausewitz, [C.] von. 1832–1834. *On War* [transl. from the German by O. J. M. Jolles]. Washington: Infantry Journal Press, 641 pp., 1950.

Clutterbuck, R. L. 1966. *Long, Long War: Counterinsurgency in Malaya and Vietnam*. New York: Praeger, 206 pp.

Connor, S. and Thomas, A. 1984. How Britain sprayed Malaya with dioxin. *New Scientist*, London, 101(1393): 6–7.

DeVita, V. T., Jr, Fisher, R. I., Johnson, R. E. and Berard, C. W. 1982. Non-Hodgkin's lymphomas. In: Holland, J. F. & Frei, E., III (eds). *Cancer Medicine*. Philadelphia: Lea & Febiger, 2465 pp.: pp. 1502–1537.

Domenico, A. di, Silano, V., Viviano, G. and Zapponi, G. 1980. Accidental release of 2,3,7,8-tetrachlorodibenzo-*p*-dioxin (TCDD) at Sèveso, Italy. V. Environmental persistence of TCDD in soil. *Ecotoxicology and Environmental Safety*, New York, 4: 339–345.

Engineers, US Army Corps of. 1972. *Herbicides and Military Operations*. Washington: US Army Corps of Engineers, Engineer Strategic Studies Group Rept No. TOPOCOM 9022300, 3 volumes (27 + [141] + 98 pp.).

Gardner, A. L. and Iverson, R. E. 1968. Effect of aerially applied malathion on an urban population. *Archives of Environmental Health*, Chicago, 16: 823–826.

Henderson, G. R. G. 1955. Whirling wings over the jungle. *Air Clues*, London, 9(8): 239–243.

Henniker, M. C. A. 1955. *Red Shadow over Malaya*. Edinburgh: Wm Blackwood, 303 pp. + 8 pl.

Hickey, G. C. 1974. *Effects of Herbicides in South Vietnam. B[11]. Perceived Effects of Herbicides Used in the Highlands of South Vietnam*. Washington: National Academy of Sciences, 23 pp.

Holden, D. 1972. Israelis admit army killed Arab crops with chemical sprays. *Sunday Times*, London, 1972 (16 Jul): 2.

---

[3] 'Ecocide' signifies the large-scale destruction of wild plants and animals and of their natural habitats, that is, the destruction of natural ecosystems. The term appears to have been coined by Arthur W. Galston (Knoll & McFadden, 1970, page 71).

23

Knoll, E. and McFadden, J. N. (eds). 1970. *War Crimes and the American Conscience.* New York: Holt, Rinehart & Winston, 208 pp.

Merck, G. W. 1946. Biological warfare. *Military Surgeon* [now *Military Medicine*], Washington, **98**: 237–242.

Merck, G. W. 1947. Peacetime benefits from biological warfare research studies. *Journal of the American Veterinary Medical Association*, Schaumburg, Ill., **110**: 213–216.

Meselson, M. S., Westing, A. H. and Constable, J. D. 1972. Background material relevant to presentations at the 1970 annual meeting of the AAAS. *US Congressional Record*, Washington, **118**: 6807–6813.

Moertel, C. G. 1982. Liver. In: Holland, J. F. & Frei, E., III (eds). *Cancer Medicine.* Philadelphia: Lea & Febiger, 2465 pp.: pp. 1774–1781.

Morton, D. L. and Eilber, F. R. 1982. Soft tissue sarcomas. In: Holland, J. F. & Frei, E., III (eds). *Cancer Medicine.* Philadelphia: Lea & Febiger, 2465 pp: pp. 2141–2157.

Murphy, J. M. *et al.* 1974. *Effects of Herbicides in South Vietnam. B[12]. Beliefs, Attitudes, and Behavior of Lowland Vietnamese.* Washington: National Academy of Sciences, [299] pp.

Norman, A. G. (ed.). 1946. Studies on plant growth-regulating substances. *Botanical Gazette*, Chicago, **107**: 475–632.

Orians, G. H. and Pfeiffer, E. W. 1970. Ecological effects of the war in Vietnam. *Science*, Washington, **168**: 544–554.

Peterson, G. E. 1967. Discovery and development of 2,4-D. *Agricultural History*, Davis, Cal., **41**: 243–253.

Rose, H. A. and Rose, S. P. R. 1972. Chemical spraying as reported by refugees from South Vietnam. *Science*, Washington, **177**: 710–712.

Rosenberg, S. A. 1982. Hodgkin's disease. In: Holland, J. F. & Frei, E., III (eds). *Cancer Medicine.* Philadelphia: Lea & Febiger, 2465 pp.: pp. 1478–1502.

Scullard, H. H. 1961. *History of the Roman World from 753 to 146 B.C.* 3rd ed. London: Methuen, 480 pp. + 4 maps.

Tarrant, R. F. and Allard, J. 1972. Arsenic levels in urine of forest workers applying silvicides. *Archives of Environmental Health*, Chicago, **24**: 277–280.

Tung, Ton That. 1973. [Primary cancer of the liver in Viet Nam.] (In French) *Chirurgie*, Paris, **99**: 427–436.

Tung, Ton That; Anh, Trinh Kim; Tuyen, Bach Quoc; Tra, Dao Xuan; and Huyen, Nguyen Xuan. 1971. Clinical effects of massive and continuous utilization of defoliants on civilians: preliminary report. In: Vien, Nguyên Khac (ed.). *Chemical Warfare.* Hanoi: Xunhasaba, 189 pp. + 8 pl. + 1 tbl.: pp. 53–83.

Westing, A. H. 1972. Herbicidal damage to Cambodia. In: Neilands, J. B. *et al. Harvest of Death: Chemical Warfare in Vietnam and Cambodia.* New York: Free Press, 304 pp.: pp. 177–205.

Westing, A. H. 1976. In: SIPRI, *Ecological Consequences of the Second Indochina War.* Stockholm: Almqvist & Wiksell, 119 pp. + 8 pl.

Westing, A. H. 1980. In: SIPRI, *Warfare in a Fragile World: Military Impact on the Human Environment.* London: Taylor & Francis, 249 pp.

Westing, A. H. 1981a. Crop destruction as a means of war. *Bulletin of the Atomic Scientists*, Chicago, **37**(2): 38–42.

Westing, A. H. 1981b. Laotian postscript. *Nature*, London, **294**: 606.

Westing, A. H. 1982a. Environmental aftermath of warfare in Viet Nam. In: *World Armaments and Disarmament, SIPRI Yearbook 1982.* London: Taylor & Francis, 518 pp.: pp. 363–389.

Westing, A. H. 1982b. Vietnam now. *Nature*, London, **298**: 114.

Ziegler, J. L. 1982. Burkitt's lymphoma. In: Holland, J. F. & Frei, E., III (eds). *Cancer Medicine.* Philadelphia: Lea & Febiger, 2465 pp.: pp. 1537–1546.

Chapter Two
Terrestrial Plant Ecology and Forestry

2.A. Terrestrial plant ecology and forestry: Symposium summary[1]

Thái Van Trùng et al.[2]

Botanical Museum and National Herbarium, Ho Chi Minh City

This summary report examines the long-term effects on inland plant ecology and forestry of the herbicides applied to South Viet Nam during the Second Indochina War of 1961–1975.

The unprecedentedly massive use of these chemicals has produced many effects on the agriculture and ecosystems of South Viet Nam, only a small part of which are currently understood. The Working Group sees the need for a large-scale co-ordinated programme to accomplish the following four objectives:

1. To establish an accurate inventory of the extent and severity of the damage and change caused by the spraying.

2. To estimate the extent of spontaneous regeneration in the sprayed forest and other ecosystems. For such work, the existence or establishment of reliable descriptions of the forests of this region must provide a necessary base of data.

3. To develop policies of land management in order to encourage such regeneration, and, where indicated, agriculture that will minimize the damage and restore the land to maximum stability and productivity.

4. To devise systems of international co-operation and aid in order to implement those beneficial policies which are beyond the financial and technological means of the Vietnamese nation.

Over the decade starting in late 1961 at least 12 per cent of the inland forests of South Viet Nam were sprayed at least once, and many were sprayed repeatedly. Some Vietnamese estimates set this figure as high as 44 per cent. The extent of permanent damage is correlated with the total herbicide dose, as judged by matching aerial photographs with military spray records. The

---

[1] Summary report of a Working Group of the International Symposium on Herbicides and Defoliants in War: the Long-term Effects on Man and Nature, Ho Chi Minh City, 13–20 January 1983 (appendix 3).
[2] The Working Group consisted of: P. S. Ashton, B. D. Baikov, E. F. Brünig, Chan Tong Yves, Duờng Hông Dât, Vu Van Dung, A. W. Galston (Rapporteur), Trân Dình Gián, Nguyên Van Hanh, Dinh Hiêp, Pham Hoàng Hô, Hoàng Hoè (Vice-chairman), I. I. Maradudin, B. A. Molski, Bui Van Ngac, Phung Trung Ngan, Y. G. Puzachenko, Nguyên Huu Quang, P. W. Richards, A. L. Takhtajan, H. Thomasius, Thái Van Trùng (Chairman), J. H. Vandermeer and J. E. Vidal.

degree of initial damage and the rate of recovery from such damage depend on many factors, including the species involved, the dosage, the total contiguous area sprayed, the terrain, and the weather patterns. Similarly, spontaneous regeneration varies widely in the affected areas; it depends mainly on the species, the size of the area affected, and the weather. The existence of a prolonged dry season in Viet Nam certainly impedes regeneration. In some areas natural regeneration has not occurred, making artificial replanting essential. In one region studied—the inland forest in the Ma Da area of north-western Dong Nai province (in the former Phuoc Long province, Military Region III, War Zone D)—regeneration has proceeded very slowly over the past decade, as judged from satellite photographs and on-the-ground studies.

Frequently the nearby availability of seeds is the critical factor determining regeneration. The regenerated forest may differ significantly from the original one in terms of economically important species. Inventories must be made of these changes.

Once an area has been sprayed with herbicides, it may be prevented from recovering through subsequent human intervention. The Working Group notes, for example, the repeated burning of the grasses and young woody cover that become established on sprayed areas such as in the Ma Da forest; and the conversion of some such areas to agriculture. Such conversions, once effected, are difficult to reverse, and such land might best be left to agriculture.

Once policies have been developed to foster recovery, laws and social practices should be developed to minimize the deleterious effects of those practices that prevent recovery.

The ecological damage produced by spraying with herbicides may also become spontaneously worse with time. For example, areas denuded of vegetation may suffer erosion or other deleterious transformations; or they may become invaded by noxious plants, such as *Imperata*, which impede restoration of the original flora. The extent of such transformation, representing possible permanent loss of forest lands, should be accurately estimated.

With regard to agriculture, a substantial area of cultivated or potentially cultivable land may have been lost as a result of the spray operations, partly because of the high concentrations of herbicides used. The problems underlying agricultural restoration or establishment in these areas require separate and intensive study in order to determine, for example, possible danger from toxic residues, effects on soil microflora, and the best crops to use.

Making recommendations for vegetational restoration in Viet Nam is difficult because the complexity of the landscape and the variety of local conditions make generalizations impractical and even counter-productive. Each separate area must be given independent analysis. Although ingenious and provocative models have been proposed to estimate productivity and performance in a forest ecosystem, it is premature to expect these models to be usefully employed in the field in Viet Nam. The Working Grup is impressed by the high quality and prodigious quantity of work accomplished by its Vietnamese colleagues under difficult conditions and with very little support.

This encourages the Working Group to urge that their research be supported in concrete terms.

Such information as is now available, admittedly fragmentary, permits the conclusion that the combined ecological, economic and social consequences of the wartime herbicide operation have been vast and will take several generations to reverse.

Recommendation

The Working Group believes that a useful approach to restoring the damaged forest resources of southern Viet Nam would be by means of a pilot scheme for a small selected area. This could be started immediately and would provide valuable experience while a large long-term scheme was being organized. The forest at Ma Da which the Working Group visited might be suitable for such an experiment if it were protected from fire. In the Ma Da area it would be possible to find: (a) areas of undamaged forest which could serve as a base of reference and as seed sources; (b) areas of herbicide-damaged forest needing to be restored to full productivity by encouraging natural regeneration or by conversion to plantations of fast-growing trees; and (c) areas of scrub and grassland which might be reforested.

The cost of the recommended pilot scheme would not be large and could perhaps be met by grants from such agencies as the United Nations Development Programme (UNDP), United Nations Environment Programme (UNEP), Food and Agriculture Organization of the United Nations (FAO), or United Nations Educational, Scientific and Cultural Organization (Unesco).[3]

---

[3] Additionally, the Working Group deemed it appropriate, as did the Symposium participants in plenary session, that international agencies adopt steps to condemn herbicidal warfare against the environment and to ban such practices from any future military operations.

2.B. Long-term changes in dense inland forest following herbicidal attack

Dinh Hiêp[1]

Institute of Forest Inventory and Planning, Hanoi

I. Introduction

Large areas of forest and agricultural lands in South Viet Nam were attacked with herbicides during the Second Indochina War, especially from 1966 to 1969. Singled out for examination here is one such example of dense inland forest in South Viet Nam.

II. Materials and methods

The study area was the almost 170 000-ha Ma Da forest reserve in north-western Dong Nai province (in the former Phuoc Long province), approximately 75 km north-north-east of Ho Chi Minh City. The basic approach was to compare the situation just prior to 1965 (i.e., prior to most of the herbicidal attacks) with that in 1973 (several or so years after the attacks) through an examination of: (*a*) US Army Pictomap No. L8020, 1:25 000, 1965; and (*b*) US National Aeronautics and Space Administration Landsat satellite image No. N.1163-02443, 2 January 1973. Examination of these documents was facilitated by magnifying lenses, stereoscopes and other photogrammetric devices. In each case, the land-type features observed were converted for comparison to maps having a scale of 1:100 000. A number of on-site inspections were carried out between 1979 and 1982, primarily in order to assist in interpreting the features obtained from the aerial photography and imagery. The following land types were distinguished: (*a*) inland forest, in turn separated into (*i*) rich and (*ii*) medium plus poor; (*b*) treeless land; and (*c*) cultivated land.

III. Results

It was found that prior to 1965 some 114 000 ha of the Ma Da forest reserve was covered by inland evergreen forest, that is, 68 per cent of the region

---

[1] Adapted by the editor from the author's Symposium presentation. See also his paper 5.B.

(table 2.B.1). Almost 15 000 ha of this largely dipterocarp forest (13 per cent of it) was especially richly stocked. The wartime destruction resulting from both herbicidal and high-explosive attacks—reduced the forested portion of the Ma Da region dramatically, there being only about 53 000 ha present in 1973, none of it richly stocked.

Table 2.B.1. Land-type distribution for the Ma Da forest reserve, Dong Nai province, pre-1965 and 1973

Land type	Pre-1965 (ha)	1973 (ha)	Change (per cent)
Inland forest	114 470	52 540	−54
Rich	14 770	0	−100
Medium plus poor	99 700	52 540	−47
Treeless land	37 460	98 970	+164
Cultivated land	16 500	16 920	+3
Total	**168 430**	**168 430**	**0**

The about 62 000 ha formerly covered by forest had not regained a forest cover by 1973. (The modest amounts of post-war planting on the one hand and felling on the other have not been distinguished in the present study; and effects of post-war burning have not been determined.) Examination of a 1981 satellite image[2] plus on-site inspection in 1982 revealed almost no natural regeneration of forest trees in the zone of destruction except in some scattered areas, especially at the margins. The destroyed area was found to have been invaded largely by such herbaceous pioneer grasses as *Imperata cylindrica* and the exotic *Pennisetum polystachyon*.

IV. Conclusion

The reduction by more than 50 per cent of the forest area in this former war zone along with the associated site degradation and economic losses will long be felt.

[2] US National Aeronautics and Space Administration (NASA) Landsat satellite image No. N.D134-052ST, 31 November 1981.

2.C. Long-term changes in dense and open inland forests following herbicidal attack

Peter S. Ashton[1]

Harvard University

I. Introduction

Tropical forests are in the process of modification or destruction world-wide, on a scale that has been estimated to be as high as 20 million hectares per year. The damage that has been caused to the inland forests of Viet Nam as a result of the Second Indochina War (Westing, 1976; 1982) therefore contributes to a broader problem. The character of wartime damage, largely by Agent Orange and primarily during the period 1966–1969, has been studied here for two forest types by inspection of a temporal series of aerial photographs reinforced by recent on-site visits.

II. Materials and methods

The primary study area was in the Ma Da forest reserve, in north-western Dong Nai province (in the former Phuoc Long province), approximately 75 km north-north-east of Ho Chi Minh City. The area has reddish-yellow podzolic soil and has been occupied by a closed-canopy seasonal evergreen dipterocarp forest. This forest—which occurs where the mean annual rainfall exceeds 1 500 mm and where there is a dry season of about 4 months—has a tall closed canopy with a woody understory; it is never penetrated by fire (Ashton *et al.*, 1978).

Aerial photographs of the Ma Da study area were available for examination from 1965 (US, 1 : 50 000 and 1 : 25 000), 1969 (US, 1 : 50 000) and 1972 (US, 1 : 20 000 and 1 : 9 320). The materials of Hiêp (paper 2.B) were also available. Information on wartime spray missions (dates and approximate locations) was available as well, from US Air Force computer-stored files. Two one-day on-site inspections of sprayed and unsprayed areas of this dense inland forest area were made in January 1983.

---

[1] Adapted by the editor from the author's Symposium presentation as revised by the author on the basis of on-site inspection.

The second study area was in the Lam Dong forest in north-eastern Dong Nai province (in the former Long Khanh province), approximately 100 km north-east of Ho Chi Minh City. The area has sandy alluvial soil and has been occupied by an open seasonal semi-evergreen to deciduous dipterocarp forest. This forest—which occurs where the mean annual rainfall is less than 1 500 mm and where there is a dry season of perhaps 4–7 months—has a medium stature and a more-or-less dense grassy ground cover. A one-day on-site inspection of sprayed and unsprayed areas of open inland forest area was made in January 1983.

III. Results

Closed evergreen (dense) inland forest

Examination of the largely pre-spray aerial photography of 1965 showed that the closed evergreen (dense) inland forest study area consisted of a patchwork of varying conditions, successional stages, and ages of dipterocarp forest and some bamboo brake. This pattern appears to have been largely, if not entirely, the result of the regional practice of shifting slash-and-burn agriculture. (Indeed, this form of agriculture was seen to be prevalent in the area in 1965; it had declined by 1969 and had ceased by 1972.) According to Hiệp (paper 2.B), prior to 1965 the Ma Da forest region was about 68 per cent forested; and that about 13 per cent of this forested portion was especially richly stocked (the 'rich' category possibly referring to primary forest). The aerial photographs of 1969 revealed numerous lengthy swaths through the forest, each approximately 500 m in width, that had resulted from the herbicidal attacks between 1965 and 1969. Somewhat more than half of the Ma Da forest was affected in this way (Hiệp, paper 2.B). Most of the swaths run in a roughly north–south direction, although some follow the major streams and roads.

In the 1969 photographs the spray damage manifested itself as changes in the albedo of the forest canopy: the crowns of the overstory had turned a uniform pale greyish white in the black-and-white photographs, with a scattering of dark (living) trees. Three years later (i.e., as seen in the 1972 photographs) the effects of the spraying had become more pronounced and the pattern of vegetational cover more heterogeneous. This variability of effect seems to have resulted from a combination of the varied natures of the site, of the original forest cover, and of the spray history (its frequency, its timing with respect to season, etc.). Another important factor must have been the post-spray fire history, as discussed below.

By 1972 large portions of the swath sites had developed a vegetational cover with an even-appearing canopy in the aerial photographs, interpreted to represent perennial grassland. The establishment and continued maintenance of this grassy cover strongly suggests the intercession of repeated post-spray fires. Bamboo brakes had become established along stream banks.

Close inspection of the forest canopy of the areas between the destroyed swaths in the 1972 photographs suggests a considerable level of mortality in the overstory trees in some of these areas, but not in others. There is no indication that any of the inter-swath areas had been subjected to burning. The observed differences appear to represent differences in frequency or intensity of direct or drift exposure to the applied herbicides (some of the swaths through the Ma Da forest having been sprayed on perhaps as many as seven occasions). Modest herbicide exposure may have killed most of the overstory, thereby permitting a relatively unaffected understory to form the new stand, whereas more severe or frequent exposure would have killed most of the trees in both overstory and understory. On the other hand, the possibility exists that most of the trees in an understory could in time succumb even if not originally killed by the herbicides, having lost the protection from sun and wind that is afforded by the normal overstory. Indeed, there are some indications that the latter was the situation in portions of the Ma Da forest.

On-site inspection in January 1983 of the grassy swaths seen in the aerial photographs revealed that the soils were mainly pale yellow-red podzolics with small laterite nodules aggregated in places near or on the surface, a normal phenomenon for this region (see below). The vegetation was dominated by the grasses *Imperata cylindrica* and *Pennisetum polystachyon*, with occasional patches of a spiny *Mimosa* about 2 m tall. Signs of recent burning were plentiful. There was an abundance of bovine and cervid hoofprints as well as some elephant dung. It appears that the ungulates are prospering on the young grass shoots that follow from burning. In a few locations there was some invasion of the grassland by pioneer woody vegetation (commercially valueless) which included *Adina sessilifolia*, *Randia tomentosa*, *Colona auriculata* and, in rare instances, the more valuable *Sindora cochinchinensis*. It was found that most of the scattered surviving relict trees noted in the various post-spray aerial photographs were *Irvingia malayana* (of little value since its wood is not durable). The bamboo that had become established along sprayed stream banks was *Oxytenanthera*.

Outside the grassy swaths (and pioneer woody vegetation) there was a forest stand with a height of only 20–35 m. Large, fully grown trees were almost or totally absent, but there was a patchy emergent canopy becoming re-established, among which *Dipterocarpus turbinatus* and *Dipterocarpus dyeri* were leading species. Below these, the main canopy was climber infested. Some commercial hardwoods appeared to be absent, for example, the formerly common *Anisoptera costata*. However, some pole-sized regeneration by species of *Dipterocarpus* and of the high-quality legumes *Sindora*, *Pahudia*, *Dalbergia* and *Pterocarpus* was observed.

A cursory search for seedling regeneration, especially in light gaps, turned up much *Shorea thorelii*, some *Sindora cochinchinensis*, but almost no *Dipterocarpus*. (*Dipterocarpus* spp. were in abundant flower but fruit production is poor, presumably owing to insect predation.) Examination of nearby unsprayed forest and experience from Kampuchea suggest that this was a normal pattern not necessarily related to the past herbicidal attacks. Moreover, the mineral soil in the unsprayed forest was similar to that in the grassy swaths, including a scattering of laterite nodules.

It is estimated that where the dense inland forest at Ma Da was sprayed less than three times, sufficient understory remained or has regenerated for a new, albeit impoverished, commercial stand to be able to reach harvestable condition within about 35 years following the spray attacks. It is further estimated that seedling establishment from the fruit of the surviving young trees should develop into a new stand comparable to the pre-spray stand after perhaps 80–100 years.

Where three or more sprayings took place, fire generally interceded and the forest was completely destroyed. (Such multiply sprayed areas that had not been subjected to subsequent burning could not be located here.) Unless fire can be excluded for long enough to permit the re-establishment of trees, reforestation is likely to prove impossible.

Open inland forest

The sprayed area examined consisted of a swath approximately 500 m in width. The more-or-less disturbed adjacent forest on either side had a stature of about 30 m with a closed semi-evergreen canopy dominated by *Dipterocarpus*, especially *Dipterocarpus intricatus*. Beneath was a dense grassy cover with sparse woody regeneration.

In the herbicide-attacked swath most trees were dead, but many trunks still stood, blackened at the base by fire to a height of perhaps 15 m. A few of the trunks had coppiced (i.e., sprouted epicormic shoots). Also, some stump sprouts of *Dipterocarpus intricatus*, *Parinari annamense* and a number of other species were being regularly harvested. The sprayed swath had been given over almost totally to the cultivation of such crops as sesame and upland rice.

IV. Conclusion

If fire prevention can be achieved for a time in the destroyed dense inland forest, reforestation could then be best achieved by artificial replanting. The simplest solution—already being carried out to some extent in the Ma Da forest—is to plant fast-growing industrial and firewood species. *Acacia auriculaeformis* is proving successful. Also worth trying in the locally prevalent poor soils are *Acacia mangium* and *Albizia lebbek*; and also perhaps species of *Pinus*, *Eucalyptus* and *Casuarina*, although the latter group would remain prone to fire damage longer than the former. A more complex approach to replanting would be to establish a fast-growing overstory (perhaps employing *Albizia falcataria* or *Samanea saman*) to be subsequently underplanted with slower-growing quality hardwoods (e.g., species of *Dipterocarpus* or *Anisoptera*). In any case, extensive single-species plantations should be avoided.

In the destroyed open inland forest it is even more necessary—and more difficult—to exclude fire if forest regeneration is to be achieved, whether by natural means (including sprouting) or by planting.

References

Ashton, P. S., Hopkins, M. J., Webb, L. J. and Williams, W. T. 1978. Natural forest: plant biology, regeneration and tree growth. In: Unesco, UNEP & FAO (eds). *Tropical Forest Ecosystems: a State-of-knowledge Report*. Paris: Unesco, 683 pp.: pp. 180–215.

Westing, A. H. 1976. In: SIPRI, *Ecological Consequences of the Second Indochina War*. Stockholm: Almqvist & Wiksell, 119 pp. + 8 pl.

Westing, A. H. 1982. Environmental aftermath of warfare in Viet Nam. In: *World Armaments and Disarmament, SIPRI Yearbook 1982*. London: Taylor & Francis, 518 pp.: pp. 363–389.

2.D. Terrestrial plant ecology and forestry: an overview

Arthur W. Galston and Paul W. Richards

Yale University and Cambridge, England

I. Introduction

Reasonably trustworthy data on the composition and quantity of herbicides sprayed in South Viet Nam during the Second Indochina War exist (Lang *et al.*, 1974; Westing, 1976). The number of times any particular forested area was sprayed is more doubtful. More questionable still are the quantitative assessments of vegetational damage, especially those for the semi-evergreen (seasonal) inland forests upon which this overview focuses. The difficulties stem in large part from a necessary heavy reliance for past information on aerial photographs, many of which are difficult to interpret owing to their poor quality, inadequate information on pre-spray conditions, and other factors. However, in the last few years some on-site information has become available. What follows must thus be viewed against a background of few known facts and much speculation.

This report begins with a review of information obtained by one of the authors towards the end of the war, more than a decade ago.[1] It then summarizes some recent unpublished data that have been collected by Vietnamese scientists.

II. Inland forests in 1970–1973

It is important to realize that, even before they were subjected to herbicide spraying for military purposes, the inland forests of South Viet Nam were very heterogeneous. Most of them were secondary and had been disturbed by shifting slash-and-burn agriculture, logging, fuel gathering, or other means. They were thus a patchwork of high forest interspersed with bamboo thickets, young regrowth, plots under cultivation and swampy areas.

The least disturbed, and economically most valuable, inland forest areas were mostly mixtures of numerous species of broadleafed trees, among which were Dipterocarpaceae (species of *Dipterocarpus*, *Hopea*, *Anisoptera* and *Shorea*) and Leguminosae (species of *Dalbergia*, *Pterocarpus*, *Sindora* and *Pahudia*). Some of

---

[1] P. W. Richards was a member of the US National Academy of Sciences Committee on the Effects of Herbicides in Vietnam (Lang *et al.*, 1974). He visited Viet Nam (and Thailand) in October 1971 and March 1972 and also had access to relevant wartime aerial photography (Richards, n.d.).

these are endemic to the region and many produce commercially important timbers. The taller (emergent) trees were 40 m or more high and had straight trunks up to 2 m in diameter at breast height.

In helicopter flights over herbicide-sprayed forest in 1971 it could be seen that almost all the emergent trees over vast areas appeared to be dead, although nearly everywhere beneath them the smaller trees and other vegetation were green and thus seemed to be alive. In short visits to forests near Dong Xoai in south-central Song Be province (in the former Phuoc Long province) which had been sprayed a year or more earlier, it was noticed that a large number of trees of *Irvingia malayana* had survived, suggesting that this particular species was relatively resistant to herbicides. Although a few seedlings and saplings of large forest trees, including one species of Dipterocarpaceae, were identified in the undergrowth of herbicide-sprayed forests in 1971, it was impossible to estimate the chances of recovery or the likely future course of ecological succession in the absence of further herbicidal or other disturbance.

The course of the secondary successions in tropical forests which have been severely damaged by logging, or which have been cleared and allowed to regenerate, as on sites abandoned after shifting slash-and-burn cultivation, is known only in its general outlines (Richards, 1952; Whitmore, 1975). Under some conditions, particularly in climates with well-marked dry seasons like that of southern Viet Nam, the succession often leads to the establishment of dense stands of the useless grass *Imperata cylindrica* or of bamboos. These stands are inflammable in the dry season and if burnt repeatedly they become more or less permanent. Conversely, if they are protected from fire, and if seed sources are available, they may in time become colonized by trees, and forest may eventually re-establish itself.

The wartime damage to the inland forests of South Viet Nam differed from that normally inflicted by lumbering or shifting slash-and-burn cultivation in two respects: first, in the enormous extent of the area affected; and second, in that the main destructive agents were herbicides, although there was also widespread damage from high-explosives. An important question therefore was whether the effects of the herbicides were fundamentally different from those resulting from other forms of disturbance in which most of the emergent trees are killed or gravely damaged. Although investigations have shown that the herbicides used during the war do not themselves persist in the soil for very long, it was possible that some of their degradation products and the contaminant dioxin may have been more lasting and have had long-continued effects on ecological succession.

III. Inland forests a decade or so later

There is evidence that war-damaged dense inland forest regions in the provinces immediately north of Ho Chi Minh City are in various stages of ecological recovery. According to recent observations by Thái Van Trùng (Botanical Museum and National Herbarium, Ho Chi Minh City, unpublished), in some herbicide-sprayed areas (as in areas where the forest has been felled) the large trees have been replaced by small fast-growing dicotyledonous species with soft light wood (e.g., *Trema, Macaranga, Mallotus*); and in others by bamboos (e.g., *Oxytenanthera, Thyrsostachys*) or herbaceous grasses (e.g., *Imperata*). Some severely war-damaged (denuded) areas have been degraded by soil erosion. Reforestation trials with teak have not so far been very successful and Trùng has proposed that it may be necessary to establish mixed multi-storied plantations. Bare land should first be protected by planting with leguminous plants (e.g., *Cassia, Indigofera, Tephrosia, Leucaena*), whose root nodules enrich the soil with nitrogen, and then underplanted with dipterocarps or other timber trees. He concluded that only a large co-ordinated effort would prevent continued deterioration of the severely war-damaged forest areas and help to restore them.

In studies of the Ma Da forest reserve in north-western Dong Nai province (in the former Phuoc Long province), approximately 75 km north-north-east of Ho Chi Minh City, Hoàng Hoè (Institute of Forest Inventory and Planning, Hanoi, unpublished) has found that in the large repeatedly sprayed areas all woody species were killed, except *Irvingia malayana* and *Parinari annamense*, and have been replaced by herbaceous tussock grasses (e.g., *Imperata cylindrica*, *Pennisetum polystachyon*) or, along streams, by bamboos. The tussock grasses, when desiccated in the dry season, greatly increased the incidence of fire. Hoè noted that natural recovery is proceeding very slowly, doing so in about 20 per cent of the herbicide-attacked area. Artificial reforestation has been carried out in this forest on a pilot basis with teak, *Eucalyptus camaldulensis* and *Hopea odorata*, but has met with only moderate success apparently because of fire and disease problems. Recently the legume *Acacia aneura* has been planted to enrich the soil prior to the planting of other trees. Although this is a costly procedure, Hoè reported that it appears to be meeting with some success.

The findings of Thái Van Trùng and Hoàng Hoè summarized above find support in the reports by Hiêp (paper 2.B) and Ashton (paper 2.C). Moreover, the present authors themselves briefly visited the Ma Da forest in January 1983 and were able to observe that some natural regeneration was now taking place in the herbicide-attacked forests, more than a decade after the spraying. Nevertheless, although the beginnings of natural recovery can be seen, it is evident that it will be many years before these forests will again approximate anything close to their original level of productivity by means of natural processes alone. It was further clear that the more severely damaged parts of the forest had been invaded by *Imperata cylindrica*, *Pennisetum polystachyon* and other grasses, and were liable to be burned during the dry seasons, a process that serves to perpetuate their presence.

IV. Conclusion

The dense semi-evergreen (seasonal) inland forests of South Viet Nam were severely damaged by herbicides and other means during the Second Indochina

War. Recent aerial photography and on-site studies suggest that early damage estimates, especially of timber losses (by, e.g., Lang *et al.*, 1974, page IV-20), may have been too low (see Lang *et al.*, 1974, page E-19). Such studies could be extended to all the forests of southern Viet Nam (and beyond), but it might be questioned whether the value of the results would justify the required heavy costs. Rather, available resources might better be allocated to studies of techniques to rehabilitate the forests.

If annual fires are permitted to continue in the grass-dominated areas, this will not only preclude the re-establishment of trees, but the grassy areas are likely to expand as well. It is thus concluded that if there is to be substantial inland forest recovery within decades rather than centuries, fires must be controlled and silvicultural techniques must be applied to accelerate the natural regeneration processes. These will include planting desirable tree species on a large scale in areas where the damage has been most severe.

There is an urgent need for a considerable amount of research on present ecological conditions in the forest areas, the best methods for encouraging natural regeneration, the selection of tree species suitable for planting, and other technical problems. Because of the great extent and diversity of the forests damaged by the wartime herbicidal attacks, and because before the war these forests had been affected to different degrees by shifting slash-and-burn agriculture, burning and other anthropogenic factors, it will probably be necessary to concentrate restoration activities in the potentially most productive forest areas. The Ma Da forest reserve appears to be a suitable site for the immediately necessary pilot studies.

Finally, the scale of the problem is so enormous, and the resources at present available to the Vietnamese government so limited, that an extensive programme of international aid appears essential.

References

Lang, A. *et al.* 1974. *Effects of Herbicides in South Vietnam. A. Summary and Conclusions.* Washington: National Academy of Sciences, [398] pp. + 8 maps.

Richards, P. W. 1952. *Tropical Rain Forest: An Ecological Study.* London: Cambridge University Press, 450 pp. + 15 pl. + 4 figs. (Reprinted with corrections 1979.)

Richards, P. W. n.d. Inland and mangrove forests of South Viet Nam in 1972–1973. *Environmental Conservation*, Geneva, in the press.

Westing, A. H. 1976. In: SIPRI, *Ecological Consequences of the Second Indochina War.* Stockholm: Almqvist & Wiksell, 119 pp. + 8 pl.

Whitmore, T. C. 1975. *Tropical Rain Forests of the Far East.* New York: Oxford University Press, 282 pp.

Chapter Three
Terrestrial Animal Ecology

3.A. Terrestrial animal ecology: Symposium summary[1]

Vo Qúy *et al.*[2]
University of Hanoi

This summary report examines the long-term effects on terrestrial animal ecology of the herbicides applied to South Viet Nam during the Second Indochina War of 1961–1975.

These chemicals were sprayed in high concentrations and over large areas of forest, damaging the forest environment and causing the death of countless animals. The Working Group primarily reviewed two presentations reporting the results of two years of study of the effects of massive herbicide spraying in the A Luói valley in Binh Tri Thien province (in the former Thua Thien province, South Viet Nam, Military Region I). Over 80 per cent of this valley had been covered by tropical forest supporting a rich fauna, but was largely degraded to grassland as a result of the spraying.

A research team led by Vo Qúy interviewed a cross-section of the inhabitants of 10 villages in the valley who had witnessed the immediate results of the spraying. These people consistently reported that spraying was followed within a few days by the deaths of large numbers of both wild and domestic birds and mammals. There has been no study investigating the contribution to this mortality from direct toxic effects of the chemicals versus indirect effects such as starvation or disease that would follow the destruction of the forest environment of animals.

An important comparison between A Luói valley and two control forest areas regarding numbers of bird species (by Vo Qúy) and numbers of mammal species (Huỳnh *et al.*, paper 3.B) was presented and discussed. Only 24 species of birds and 5 species of mammals were found in A Luói valley whereas 145 and 170 bird species and 30 and 55 mammal species were recorded in the two control forests.

Two additional studies were reported. According to L. W. Medvedev, termite abundances were lower in a sprayed forest site as compared with an unsprayed forest site of similar vegetational structure. According to Tran The Thong, higher incidences of reproductive problems and birth abnormalities were found among

[1] Summary report of a Working Group of the International Symposium on Herbicides and Defoliants in War: the Long-term Effects on Man and Nature, Ho Chi Minh City, 13–20 January 1983 (appendix 3).
[2] The Working Group consisted of: Dang Huy Huỳnh (Vice-chairman), V. Landa, M. Leighton (Rapporteur), L. W. Medvedev, I. Mototani, E. W. Pfeiffer, Vo Qúy (Chairman), V. E. Sokolov and Tran The Thong.

domestic pigs in a village that had been subjected to chemical spraying as compared with a similar unsprayed village.

Visits to defoliated forests and examination of aerial photographs of sprayed and unsprayed forest have shown that the tropical forest was transformed by the spraying to one of two types of degraded vegetation. First, forest repeatedly or intensely defoliated over large areas was often subsequently burned, leading to the establishment of grassland. Examples are the A Luói valley mentioned above and the large inland forests in and north of the Ma Da area of north-western Dong Nai province (in the former Phuoc Long province, South Viet Nam, Military Region III, War Zone D). Second, over large areas of forest less frequently sprayed, plants of the upper layers (strata) of the forest were killed, resulting in a new forest of low stature relatively poor in animal species.

Thus in defoliated areas, tropical forest supporting a rich fauna of invertebrates and vertebrates has been destroyed, together with the animals dependent upon the microclimatic conditions, food resources and physical structure of the forest. Populations of animals requiring forest of well-developed structure and high plant species diversity have been reduced and subdivided into isolated areas. These species are now more susceptible to local extirpation as a result of the reduction and division of their forest habitat. This phenomenon was specifically investigated by Vo Qúy and colleagues in seven forest areas of southern Viet Nam during surveys of endangered species such as the Javan rhinoceros, kouprey, douc langur, and Edwards's pheasant, and of other economically important vertebrates.

Recommendations

The Working Group suggests the following objectives for further research on the ecological impact of chemical warfare with herbicides on forest animals:

1. Thorough ecological and zoological studies are necessary, especially in order to quantitatively document differences in animal species richness and abundance in sprayed and unsprayed areas for different forest types.

2. Ecological field studies should be combined with laboratory investigations of particular animal taxa in order to discover species useful as bio-indicators of herbicidal impact. Similar studies should be used to investigate whether long-term reproductive problems have resulted from genetic damage to wild and domestic animals surviving chemical poisoning.

3. The distribution of any residual chemicals in the ecosystem should be assessed.

4. Long-term research plots in forest areas recovering from chemical spraying should be established in order to monitor changes in their animal communities.

5. More surveys should be conducted in order to identify and categorize the remaining forests of southern Viet Nam and their animal components.

The Working Group stresses that recommendations from animal ecologists for forest rehabilitation must be integrated with economic studies of how best to utilize these altered lands for the economic and social needs of the people. The Working Group has two immediate recommendations to offer:

1. It is suggested that a system of national biological reserves be established in order to protect and manage what remains of the rich diversity of animal life in the forests of southern Viet Nam.

2. There is a special concern about further reductions in forest cover caused by the spread of grasslands, this process having been set in motion by the herbicidal attacks. It is thus suggested that efforts be devoted to reforesting grassland in order to rejoin small patches of forest that are now isolated from one another, a situation which prevents animal dispersal.

Finally, the Working Group suggests that biological institutions within Viet Nam seek expert assistance and funds from international agencies such as the Food and Agriculture Organization of the United Nations (FAO), the United Nations Development Programme (UNDP), the United Nations Environment Programme (UNEP), the United Nations Educational, Scientific and Cultural Organization (Unesco) and especially its Man and the Biosphere Programme (MAB), the International Union for Conservation of Nature and Natural Resources (IUCN), and the World Wildlife Fund (WWF).

In closing, the Working Group wishes to emphasize that the complexity of and interrelationships among the ecological problems that have been identified require co-operation among botanists, zoologists, soil scientists and aquatic biologists in order to achieve the rehabilitation of the devastated forest fauna.

3.B. Long-term changes in the mammalian fauna following herbicidal attack

Dang Huy Huỳnh, Dan Ngoc Cân, Quôc Anh and Nguyên Van Tháng[1]

National Center for Scientific Research, Hanoi

I. Introduction

Large areas of forest and agricultural lands in South Viet Nam were attacked with herbicides during the Second Indochina War, especially from 1966 to 1969. Long-term effects on the indigenous mammalian fauna can result from the the indirect influence of disturbing the vegetation (loss of food or cover) as well as from the direct toxic action of the applied chemicals. In the present study changes in the mammalian fauna have been investigated for an inland (upland) forest habitat more than a decade following the herbicidal attacks.

II. Materials and methods

The previously sprayed study area was the A Luói valley, a large valley primarily in A Luói district (and extending somewhat into Huong Hóa district), Binh Tri Thien province (in the former Thua Thien province). The valley is on the eastern slopes of the Truong Son mountain range, bordering Laos and roughly 50 km south of Quang Tri City and 100 km west of Danang. The A Luói valley is about 124 000 ha in size of which about 107 000 ha, or 86 per cent, is forest land and much of the remainder grassland.

Prior to the spray attacks the A Luói valley supported a luxuriant forest flora, much of it primary forest, including several highly valuable timber species. The herbicide-destroyed areas—representing a substantial fraction of the forest land—now support the grass *Imperata cylindrica* as well as a number of low shrubs and vines, but almost no timber trees.

The unsprayed comparison study site was in the south-eastern portion of the A Luói valley and was chosen for its geographical and habitat comparability to the sprayed site.

The mammal census was carried out during 1981–82 by visual means (using field glasses and cameras) and via collections (using nets, traps, guns and

---

[1] Adapted by the editor from the authors' Symposium presentation.

49

cross-bows). Observations were carried out during the periods 0600–1000, 1400–1800 and 2000–2400 hours, whereas the trapping was carried out continuously for a period of seven days. Positive indentification of the collected specimens was aided by dissection as necessary. Further information on the local fauna was obtained from the published literature and by interviewing local inhabitants.

III. Results

Livestock

According to A Luói district officials, a high proportion of the local domestic water buffaloes and cattle (zebus) died during the period of heavy spraying in 1968–1969. For example, of some 5 000 such livestock in Cam Lô village only about 170 were reported to have survived. Pigs were also affected to some extent.

Wildlife

The region in question was once rich in wild mammals and used to be considered a paradise for hunters of large game animals (Thao, 1976). A partial listing of wild mammals indigenous to the A Luói valley is presented in table 3.B.1.

Local inhabitants reported having found dead wildlife at the time of the spraying (i.e., two or three days following it), including deer, serows, wild boars, douc langurs and white-cheeked gibbons.

Now, more than a decade later, numerous mammals once common to the sprayed area have become rare or very rare, including, especially, wild boars, wild water buffaloes, deer, serows and tigers (table 3.B.1). Moreover, several of the mammals now rare there remain present in the comparable unsprayed area. Conversely, the fawn-coloured mouse and Sladen's and other rats, once very rare, have become common to the sprayed sites.

IV. Conclusion

Although some of the observed wildlife losses may have had their basis in direct poisoning at the time of spraying, the major cause has been the disappearance of the required habitat: the loss of food and of cover (see also Westing, 1976, page 32). The increase in small rodents is also based on the changed vegetational structure.

The herbicidal destruction of forest habitats has been a major contributor to the war-caused disruption of wildlife in Viet Nam and has exacerbated the precarious status of a number of endangered species, including that of the wild water buffalo, Asian elephant, gaur, white-cheeked gibbon, douc langur, leopard, clouded leopard and Javan rhinoceros (Constable, 1981–1982; Gochfeld, 1975; Ngan, 1968; Nowak, 1976; Westing & Westing, 1981).

Table 3.B.1. Wild mammals indigenous to A Luói district, Binh Tri Thien province, 1981–1982 (partial listing)

Common name[a]	Scientific name[a]	Pre-spray status in sprayed site[b]	Present status in sprayed site	Present status in unsprayed site
Bear, Asiatic black	*Selenarctos thibetanus*	Present	Rare	Present
Bear, Malayan sun	*Helarctos malayanus*	Present	Rare	
Boar, wild	*Sus scrofa*	Common	Rare	Present
Buffalo, wild water	*Bubalus bubalis*	Common	Very rare	
Cat, leopard	*Felis bengalensis*	Present	Rare	Present
Civet, large	*Viverra zibetha*	Present	Very rare	
Civet, small	*Viverricula indica*	Present	Rare	
Deer, lesser mouse	*Tragulus javanicus*	Present	Rare	
Deer (as a group)[c]	Family Cervidae	Common	Rare	Present
Elephant, Asian	*Elephas maximus*	Present	Very rare	
Gaur	*Bos gaurus*	Present	Very rare	
Gibbon, white-cheeked	*Hylobates concolor*	Present	Very rare	Present
Langur, douc	*Pygathrix nemaeus*	Present	Very rare	
Leopard	*Panthera pardus*	Present	Absent?	
Leopard, clouded	*Neofelis nebulosa*	Present	Absent?	
Macaques	*Macaca* spp.	Present	Rare	Present
Mouse, fawn-coloured	*Mus cervicolor*	Absent?	Common	
Porcupine, brush-tailed	*Atherurus macrourus*	Present	Rare	
Porcupine, Chinese	*Hystrix hodgsoni*	Present	Rare	
Rat, Sladen's	*Rattus sladeni*	Absent?	Common	
Rats	*Rattus* spp.	Rare	Common	
Rhinoceros, Javan	*Rhinoceros sondaicus*	Present	Absent?	
Sambar (barking deer)	*Cervus unicolor*	Present	Rare	
Serow (Sumatran goat)	*Capricornis sumatraensis*	Common	Very rare	
Squirrel, giant flying	*Petaurista petaurista*	Present	Very rare	Present
Squirrels (as a group)[d]	Family Sciuridae	Present	Rare	Present
Tiger	*Panthera tigris*	Common	Rare	

Sources and notes:

[a] Nomenclature is according to Van Peenen (1969).
[b] From Chochod (1950), Morice (1875), Osgood (1932), Thao (1976), and local interviews.
[c] The deer here include the muntjac (*Muntiacus muntjac*), sambar (*Cervus unicolor*), and others.
[d] The squirrels here include the giant flying squirrel (*Petaurista petaurista*), striped ground squirrel (*Menetes berdmorei*), striped tree squirrel (*Tamiops* sp.), tree squirrel (*Callosciurus erythraeus*), and others.

In order to ameliorate the situation in the previously sprayed areas it is recommended that the cutting of wood and the hunting of game be prohibited there. The existing reforestation programme should also be accelerated. For the reclamation of severely eroded slopes the planting of *Artocarpus* (jack fruit) is recommended, both for its soil-holding ability and its value as browse for wildlife. The problems associated with the elevated rodent populations in the sprayed areas (human diseases, crop losses) must also be addressed.

References

Chochod, L. 1950. *La Faune Indochinoise: Vingt-cinq Années de Chasses au Tonkin et en Annam.* Paris: Payot, 200 pp.

Constable, J. D. 1981–1982. Visit to Vietnam. *Oryx*, London, **16**: 249–254.

Gochfeld, M. 1975. Other victims of the Vietnam war. *BioScience*, Washington, **25**: 540–541.

Morice, A. 1875. *Coup d'Oeil sur la Faune de la Cochinchine Française.* Lyon: H. Georg, 101 pp.

Ngan, Phung Trung. 1968. Status of conservation in South Vietnam. In: Talbot, L. M. & Talbot, M. H. (eds). *Conservation in Tropical South East Asia.* Morges [now Gland], Switzerland: IUCN Publ. N.S. No. 10, 550 pp.: pp. 519–522.

Nowak, R. M. 1976. Wildlife of Indochina: tragedy or opportunity? *National Parks & Conservation Journal*, Washington, **50**(6): 13–18.

Osgood, W. H. 1932. Mammals of the Kelley-Roosevelts and Delacour Asiatic expeditions. *Publications of Field Museum of Natural History Zoological Series*, Chicago, **18**(10): 191–339 + pl. IX.

Thao, Lê Bá. 1976. *Thiên Nhiên Viêt Nam [Nature in Viet Nam].* Hanoi: Nhà Xuât Ban KHKT [KHKT Scientific & Technical Publishing House].

Van Peenen, P. F. D. 1969. *Preliminary Identification Manual for Mammals of South Vietnam.* Washington: US National Museum, 310 pp.

Westing, A. H. 1976. In: SIPRI, *Ecological Consequences of the Second Indochina War.* Stockholm: Almqvist & Wiksell, 119 pp. + 8 pl.

Westing, A. H. and Westing, C. E. 1981. Endangered species and habitats of Viet Nam. *Environmental Conservation*, Geneva, **8**: 59–62.

3.C. Terrestrial animal ecology: an overview

Mark Leighton

Harvard University

I. Introduction

Perhaps in no other aspect of the long-term consequences of chemical warfare in Viet Nam are we as ignorant as in the effects on the animal component of the inland forest. This ignorance stems from the limited effort that zoologists and ecologists have been able to direct towards taking censuses of the sprayed forest during or after the Second Indochina War, and from the more extensive and time-consuming sampling methods necessary for collection of even simple presence or absence data for animals, especially invertebrates, in contrast to plants. Whereas the forestry industry in Viet Nam prior to the war generated at least some baseline data on the plant species composition of inland forests, no really similar economic stimulus for study was provided by the forest fauna. Furthermore, although aerial photography can document temporal changes in vegetation and even in species composition over large areas, no similarly efficient method is available for recording animals, even for the larger vertebrates.

Given the absence of faunal inventories prior to spraying, only indirect, retrospective studies of changes in faunal composition are now possible. In such studies, sites in sprayed and unsprayed areas must be matched for factors affecting population densities and species composition other than chemical spraying. This engenders a daunting list of factors for an ecologist to define and control, given the multitude of abiotic and biotic factors that define the niche of each tropical forest animal (see, e.g., Gilbert, 1980) and our naivety about the causes of local site-to-site variation in animal abundance within even an apparently homogeneous habitat. Hence it would be difficult to maintain that specific population densities after the fact are the result of chemical application *per se*, rather than from the habitat transformations that have occurred following attack by herbicides.

The immediate direct effects on animals of the application of herbicides to the inland forests of South Viet Nam are essentially undocumented. Field trials with various animal taxa indicate that the components of Agent Orange are poisonous to vertebrate and invertebrate life, as applied at military concentrations. There is, however, little evidence for residual effects of these chemicals on the health of animal populations beyond a few weeks, months or years, depending on the organism (see, e.g., Westing, 1976).

It is important to emphasize in this regard, that debating the extent to which various elements of the fauna were killed immediately by direct chemical poisoning is superfluous to our deliberations about the long-term effects of the chemicals on animal ecology. Destruction of forest habitat, as occurs when the forest is transformed to a degraded vegetation by herbicidal attack, is accompanied by the death of the fauna dependent on the food resources, microclimate, and other conditions prevailing in a forest. In the case of chemical warfare, the fauna not directly killed but nonetheless forced to emigrate because of habitat degradation will in turn perish in large part, because with few exceptions the adjacent unsprayed forest is at its carrying capacity for these animal populations. Hence, lack of direct evidence for toxic effects of defoliants should not delude one into imagining that the individual animals living in a sprayed forest were able to disperse and thereby survive the assault on their habitat.

The close association between an animal's distribution and its particular habitat is a fundamental rule of animal ecology (Elton, 1950; Ricklefs, 1979). It necessarily follows that sufficiently disruptive alteration of a habitat leads to replacement of its fauna by species more adapted to the altered conditions that come to prevail, as long as these new colonists are able to disperse into the habitat. That herbicide spraying has indeed dramatically transformed good habitats into degraded vegetation types is clear from post-spray aerial photography and on-site inspection (Lang *et al.*, 1974; Orians & Pfeiffer, 1970; Westing, 1976, 1982; also personal observation in 1983). The grasslands of tropical Asia, largely anthropogenic in origin, are notably poor in species (Medway & Wells, 1976). This is unlike temperate or sub-tropical grasslands, which have a long history that has allowed evolutionary adaptation to them. These inferences from ecology emphasize the need to assess what sorts of vegetative succession have occurred in the more than a decade since spraying, what types of habitat the transformed vegetation now provide for animals, and what the prognosis is for habitat improvement, through further natural ecological succession.

II. Field studies of animal ecology in defoliated forest

There is no literature available for Viet Nam on the indirect long-term effects of Agent Orange or the other military herbicides on animal ecology. Understandably, research efforts within Viet Nam since the end of the war have been directed towards evaluating the toxic effects of herbicides on humans and towards rehabilitating destroyed forest (especially the mangroves) for commercial wood production.

Huỳnh *et al.* (paper 3.B) have studied the mammalian fauna of forest land in the A Lưới valley (A Lưới district, Bình Trị Thiên province) west of Danang near the Laotian border, contrasting an unsprayed portion with a comparable portion that had been destroyed by herbicides more than a decade previously; Vo Qúy (University of Hanoi, unpublished) has done the same for the avian fauna. The vegetation of the A Lưới valley was formerly comprised mostly of forest but was largely reduced to and remains a grassland after multiple episodes of chemical spraying and subsequent burning. Not unexpectedly, the vertebrate community was much poorer in species following forest conversion to grassland. These investigators also carried out regional surveys to define the current distributions of endangered mammals and birds, especially of primates, bovids, rhinoceroses and pheasants. Although it is clear that these species have become further restricted in distribution following the war, it is impossible to distinguish the direct effects of chemical (and high-explosive) attack from the resulting indirect effects of habitat destruction. Extensive elimination of forests caused by defoliation would in any case largely preclude reoccupation by these species (see also Constable, 1981–1982; Westing & Westing, 1981).

III. Effects on animal ecology as inferred from habitat degradation

It is possible to extend the information from the rather meagre field studies of changes in animal distribution and abundance by drawing on what is known from elsewhere of associations between faunal composition and vegetation in tropical forests, and by discussing these findings in light of the vegetational changes that followed spraying, the vegetational types that now prevail, and the probable dynamics of this vegetation in the future.

It should first be emphasized that herbicides were disproportionately sprayed on the 'dense' forest of South Viet Nam (Lang *et al.*, 1974; Westing, 1976). At a conservative estimate, 14 per cent of the 10.4 million hectares of the woody vegetation of South Viet Nam was defoliated; in contrast, 19 per cent of the 5.8 million hectares of dense forest was sprayed (Westing, table 1.6). This is an important point to bear in mind, because tropical dense forest (or tropical semi-evergreen rain forest, as classified by Whitmore [1975]) is overwhelmingly the most species-rich of the inland forest types of Viet Nam, a richness extending across animal phyla. The dense forest provides the greatest habitat diversity of any of the Indochinese forests, owing to its tall structure, diversity of plant structural forms, plant species richness, reduced seasonality and greater equability of microclimatic conditions. A great proportion of the fauna of such forests in turn tends to be more stenotypic (i.e. specialized as to particular habitat requirements) and less buffered against vegetative change than are the faunas of forests where natural disturbances (fire, drought, and so on) are a more common occurrence.

It is useful to classify sprayed areas that were formerly forested as either secondary (successional) forest or as open grassland. In fact, each vegetational type undoubtedly shows variability depending on the timing and frequency of spraying and on local site conditions. This crude dichotomy must suffice, however, given the absence of further study which might lead to a more detailed classification. This crudeness is not a major objection, however, because these two types of vegetation are very different, as are their prospects for natural habitat recovery. The secondary forest is comprised of a woody vegetation low in stature, without the vertical layering that characterizes primary forest or late secondary forest. In some cases it largely consists of juvenile trees that were in the understory when the canopy trees were killed by herbicides. Consequently, this vegetation might be expected to have a relatively low diversity of plant species and of microhabitat conditions, and to be particularly poor in mature (reproductively active) plants. Hence, this forest may be unable to support the divers fauna of nectar, pollen, seed and fruit eaters characteristic of a forest of greater diversity. This contrasts with other secondary forests in which regrowth proceeds by successive colonization and maturation of plants of differing longevities, so that floral and fruit resources are at greater densities during the period of forest re-establishment (Whitmore, 1975).

Open grasslands became established when repeated defoliation was followed by burning of the accumulated dry litter. Fires sometimes were initiated as military policy (Westing, 1976, page 58), were accidentally or deliberately set by agriculturalists or hunters, or perhaps started by lightning strikes. Subsequent initiation of woody succession from this grassland has been arrested, apparently by periodic reburning during the dry season.

There are four important questions that need to be addressed in order to assess the effects on forest fauna of this habitat degradation to either grassland or early secondary forest:

1. What are the current locations and total areas of different types of vegetation that have replaced sprayed forest? It is especially important to distinguish between secondary forest and grassland, since the woody growth—but not the grassland—is presumably returning slowly to a richer forest habitat. Mapping the distribution of sprayed areas is crucial for identifying areas of potential for nature reserves or for rehabilitation of habitat and wildlife, while at the same time excluding, at least for the moment, areas that were too extensively damaged or dissected to serve such purposes.

2. In what spatial pattern are areas of degraded forest intermingled with undamaged forest? Knowledge of the sizes of patches of good forest, the distances between patches of any one type and the quality of habitat of the intervening vegetation separating habitable forest is essential in order to evaluate the likelihood that so-called island effects might restrict persistence or recolonization of populations in otherwise acceptable habitat (see, e.g., Diamond, 1976; Frankel & Soulé, 1981). Forest separated by grassland, by low woody vegetation, or by agricultural plots may be essentially inaccessible to some animals for recolonization. This is so because the animals lack the physical abilities to disperse or are restricted by psychological barriers from crossing the inappropriate habitat. Moreover, species that remain after spraying or are able to recolonize may nonetheless be unable to persist in the remnants of good forest, if the forest occurs as patches insufficient in size or resource diversity to maintain a viable population. This is especially a problem for animals that have specialized habitat requirements or require large home ranges. Insects fitting this description include trap-lining (euglossine) bees or moths that are specialized feeders on scattered plants (Gilbert, 1980). Vertebrates include those taxa whose food supplies are

sparsely distributed or fluctuate seasonally so that they must shift feeding areas in response to local changes in resource supply. Most frugivorous and carnivorous birds and mammals fit into one or the other of these categories (Leighton & Leighton, 1983; Terborgh & Winter, 1980).

3. What faunal composition is currently associated with each type of unsprayed forest of Viet Nam and with the comparable sprayed site? An analysis of the severity of defoliation on animal communities and the prognosis for the recovery of these communities to their former richness depends on recording which populations are present in contemporary forests. Through such data generated on habitat associations, it is possible to make a guess as to whether a species is absent because the appropriate habitat is unavailable or else because of local extirpation for some other reason. An absent species might recognize or be reintroduced to the appropriate habitat, if the appropriate habitat were currently available or once it became reconstituted through natural succession or via human manipulation. Thus the prognosis for natural recolonization and population establishment versus managed reintroduction, perhaps in conjunction with habitat improvement, is important to evaluate. A number of taxa deserve special attention in these censuses: (a) endangered species, including invertebrate taxa, the latter often overlooked by conservationists; (b) species important in nutrient recycling such as termites and ants; and (c) species known to service plants as seed dispersal agents (many bird and mammalian taxa; see, e.g., Leighton & Leighton [1983]) or pollinators (various insect taxa, including groups of thrips, bees, flies and beetles, as well as some taxa of bats and of birds; see, e.g., Whitmore [1975]).

It is axiomatic that extreme interdependence characterizes the associations among a particular animal and other elements of the biota in tropical forests. Complex interactions thus link the population density of one species to that of others in these communities. These linkages are at present undeciphered for the great majority of tropical species. Hence, we cannot make sound predictions about the faunal changes that have occurred following herbicidal decimation, and that will continue to occur in conjunction with the dynamics of vegetational change, without further study.

4. What is the prognosis for further habitat change in sprayed areas? This depends first of all on whether such change is the result of natural succession or whether it is anthropogenic, especially so in the latter instance if the conversion of vegetation is to agriculture. The critical problem that needs to be addressed for the grassland areas is the role that fire plays, and will continue to play, in retarding forest recovery. Are grasslands progressively encroaching on forest as fire kills the woody vegetation along its margin, or vice versa? This can be most satisfactorily answered by comparing a succession of aerial photographs or Landsat imagery for areas in which the recent history of land use is known. An ancillary question concerns the rate at which low, secondary forest which has been generated by spraying is proceeding towards the more layered, taller forest capable of supporting a diverse fauna. The answers to these questions relate to the degree to which natural versus managed regeneration is necessary to improve forest habitat.

In sum, degradation in structure, in habitat complexity, in plant species richness, and in climatic buffering have all persisted in the inland forests as a legacy of chemical spraying. Although the degraded forests are now undergoing a slow succession towards tall, species-rich forest, this replacement vegetation undoubtedly remains highly depauperate compared with that from which it was derived. And it appears that many of the grassland patches have not commenced a woody succession at all.

IV. Recommendations

Further projects by animal ecologists should be directed first towards developing a land-use policy for the inland forests, and second towards research designed to evaluate the dynamics of faunal change in forests subjected to chemical disruption.

To evaluate the potential of different areas for conservation, adequate faunal inventories of the different types of vegetation must be produced, the pattern of existing vegetation must be mapped and the expected course of future changes in plant and animal communities must be assessed. Whereas Vietnamese scientists are capable of tackling the tasks required to fulfil these aims, severe labour, equipment and funding shortages indicate that aid from international scientific agencies is necessary.

The role of animal ecologists in formulating land-use policy

From the earliest stages of policy formation it would be most efficient if the research and conservation objectives of animal ecologists were to be developed in consultation with plant ecologists, foresters, agricultural policy makers and others. Through such co-operation, those areas which should be earmarked for commercial production can be distinguished from those which might well be set aside as biological reserves or as parks for conservation, research, or recreation. Consultation with other scientists would enable the identification of areas both in need of faunal surveys and available for long-term ecological studies, which, of course, should be integrated with the researches proposed by botanists, foresters and soil scientists.

Many of the sprayed and unsprayed forest areas of southern Viet Nam require faunal inventories in order to rank their usefulness as biological reserves. Up-to-date aerial photography is certainly required for some areas in which the present mosaic of vegetational formations is incompletely known. Once the forested areas are mapped, potential sites for nature reserves can be selected for ground survey. This system of reserves should be established on the basis of at least the following criteria:

1. Reserves should be of a size sufficient to maintain biological populations. It is impractical to suggest that habitat necessary to support 500 or more individuals of a population (a rule-of-thumb magnitude that minimizes the chance

of extinction [Frankel & Soulé, 1981]) can be protected for all species. Rare large mammals such as the kouprey, Sumatran rhinoceros or Javan rhinoceros, each apparently numbering less than two dozen individuals in Viet Nam, are on the edge of extinction (Constable, 1981–1982; Gochfeld, 1975; Nowak, 1976; Westing & Westing, 1981). The best course to follow in such cases—given that preservation of species such as these is a priority of the government of Viet Nam—is to include their known centres of distribution within nature reserves and hope that with assiduous management the populations will recover.

2. Reserves should be selected to represent the diversity of habitats important to animal communities. Recognizing the principle that different vegetational formations are associated with distinct animal communities, forests representing the diversity of altitudes, soils and climatic regimes of southern Viet Nam should each be preserved.

3. Conservation compromises will be necessary, even when the only interests under consideration are those of biologists. For example, a forest of special interest, particularly rich in animal and plant species or with important endangered species, may have been heavily degraded by spraying or reduced in area. Such a forest may nonetheless warrant saving over a more pristine or larger area that is of less intrinsic biological importance.

4. Biological conservation is in many cases compatible with other uses of an area, if the scale of such disturbances is limited and the habitat is permitted to recover between uses. Selective logging of hardwoods or hunting by shifting slash-and-burn agriculturalists, if of low intensity and regulated, can be allowed in zones surrounding totally protected areas, as such multi-use forests act as buffers from unregulated human disturbance.

5. Owing in part to the legacy of the Second Indochina War, intervention may be required to rehabilitate some animal habitats. If forests—favourable in other ways as nature reserves—are dissected by patches of grassland, then studies and appropriate actions should be pursued to rehabilitate these grasslands by encouraging them to commence a successional sere towards forest. In such grassland/forest mosaics, the intervening forest strips or islands are reduced in their capacity to maintain animal populations because of their small size and distance from one another. Moreover, it is of grave concern if substantial dissection by grassland prevails in many of the richer evergreen forests, or if the grasslands are expanding.

6. The widespread hunting of forest vertebrates must be controlled and confined to designated areas. Forest workers are now openly hunting with military rifles, both for deer, using headlamps at night, and for other vertebrates, during the day. Canopy vertebrates such as primates and large birds are especially at risk from the skilled riflemen and ubiquitous sophisticated arms lingering from the war. For example, such hunting has undoubtedly contributed to the present absence of primates from the Ma Da forest reserve in north-western Dong Nai province (in the former Phuoc Long province), approximately 75 km north-north-east of Ho Chi Minh City—a forest that once supported five primate species. Where animals have been eliminated from acceptable habitat by herbicidal spraying, hunting, or a combination of factors,

pairs or social groups might well be re-introduced to unoccupied habitat in order to re-establish populations.

7. Habitat improvement might be attempted by selectively encouraging increases in food density or by alleviating the effects of factors that limit the population density of the species in question (e.g., too many predators, too few nesting sites). This is certainly a risky and costly management tactic, given the potential for unforeseen ecological ramifications of human alteration of the relative abundances of organisms within a community.

Further research for animal ecologists

In addition to studies related to the objectives of land use planning noted above, a research programme ought to be undertaken designed to interpret the changes in faunal composition produced by chemical spraying and to chart the course of future changes. Again, it is essential to plan, conduct, and interpret these studies with the full participation of scientists other than animal ecologists. The studies of the A Luói valley by Vietnamese biologists and soil scientists is a commendable model.

The first step to be taken would be a study of aerial photographs to identify appropriate research sites. It is essential that the land-use and spraying history of these sites be known from before spraying until the present, therefore requiring a temporal sequence of photographs. One temporal series of photographs has been developed by Ashton (paper 2.C). It is necessary to obtain more recent aerial photography for some of these areas, but the ones now in hand, together with recent ground observations by this author and Ashton, are sufficient to suggest that the Ma Da forest reserve serve as one of these sites. The following list of advantages in using this reserve can serve to guide the selection of other research sites:

1. Excellent spraying records are available for this region, showing different intensities of spraying. The pre-spray photographs from the 1950s can be used to identify the specific types of vegetation prior to spraying within the forest; since shifting slash-and-burn agriculture was practised here, the forest was a mosaic of different stages of succession.

2. In many areas of the reserve, agriculture has been excluded since the time of spraying, so that vegetational transformations following spraying have not been confounded by agricultural practices.

3. Forest of relatively homogeneous structure has been transformed to several types of vegetation, according to the intensity of chemical spraying. In juxtaposition now are relatively unaltered tall forest, low woody regenerating forest, grassland invaded by woody colonizing species and grassland being maintained by fire in a sub-climax stage.

4. This mosaic provides an excellent set of comparisons for measuring associations between faunal composition and vegetation, inasmuch as the faunal compositions, soils and microclimate of the elements of the mosaic were similar prior to spraying.

5. This situation also presents an opportunity to study: (*a*) the role of fire in the spread or recession of grassland; and (*b*) the dynamics of faunal turnover as the vegetation changes.

6. Of special importance, the Ma Da reserve is quite accessible from Ho Chi Minh City. Moreover, in the event of petrol or vehicle shortages, study plots could be approached from the forestry headquarters within the reserve by bicycle or by foot.

Resources permitting, representative areas of other sprayed forest formations in southern Viet Nam could be selected as research sites on the basis of the Ma Da model. The mountainous forests along the border with Laos, for example, are important for wildlife and were heavily sprayed.

A fundamental objective of further ecological studies must be to record accurately different forest types for selected animal taxa. Beyond information on densities and distributions of species, next most useful would be demographic data on the age and sex compositions of their populations (Frankel & Soulé, 1981).

Once the discrete habitats for which censuses must be carried out have been defined on the basis of current vegetation and of spraying and land-use history, then stratified random sampling techniques could be adopted. Within each plot or section of each habitat, line transects or plots should be randomly selected, and permanently demarcated for subsequent periodic resampling. Decisions about the sizes and numbers of plots or transects depend, of course, on the types of organism recorded and on the precision of estimates desired by the investigator.

Differences between unsprayed forest of a particular structure, plant species composition and history, and comparable sprayed forest are to be established by comparisons of faunal composition. The ecological significance of such differences must be interpreted in relation to the roles that various taxa play in the ecosystem (Ashton *et al.*, 1978; Gilbert, 1980). In consequence, faunal inventories should emphasize those species important in nutrient recycling (Herrera *et al.*, 1978; Whitmore, 1975); those active in the biological control of other populations (Raven *et al.*, 1980); those that are required by plants as pollinating agents (Janzen, 1975); and those necessary as agents of seed dispersal (Leighton & Leighton, 1983). The study plots established in sprayed and unsprayed forest can also serve well as sites for other research in Vietnamese tropical biology.

In closing, emphasizing once again our ignorance of the consequences of large-scale chemical spraying for tropical animal communities, it is strongly recommended that broadly-based ecological studies of Vietnamese forests be carried out. Although we have but limited knowledge of the ecology of tropical animals, we are fully cognizant of the fragility of tropical ecosystems and their inherent susceptibility to environmental assault.

References

Ashton, P. S., Hopkins, M. J., Webb, L. J. and Williams, W. T. 1978. Natural forest: plant biology, regeneration and tree growth. In: Unesco, UNEP, & FAO (eds). *Tropical Forest Ecosystems: a State-of-knowledge Report*. Paris: Unesco, 683 pp.: pp. 180–215.

Constable, J. D. 1981–1982. Visit to Vietnam. *Oryx*, London, **16**: 249–254.

Diamond, J. M. 1976. Island biogeography and conservation: strategy and limitations. *Science*, Washington, **193**: 1027–1029.

Elton, C. 1950. *Ecology of Animals*. 3rd ed. London: Methuen, 97 pp.

Frankel, O. H. and Soulé, M. E. 1981. *Conservation and Evolution*. Cambridge, England: Cambridge University Press, 327 pp.

Gilbert, L. E. 1980. Food web organization and the conservation of neotropical diversity. In: Soulé, M. E. & Wilcox, B. A. (eds). *Conservation Biology: an Evolutionary-ecological Perspective*. Sunderland, Mass.: Sinauer Associates, 395 pp.: pp. 11–33.

Gochfeld, M. 1975. Other victims of the Vietnam war. *BioScience*, Washington, **25**: 540–541.

Herrera, R., Jordan, C. F., Klinge, H. and Medina, E. 1978. Amazon ecosystems: their structure and functioning with particular emphasis on nutrients. *Interciencia*, Caracas, **3**: 223–232.

Janzen, D. H. 1975. *Ecology of Plants in the Tropics*. London: Edward Arnold, 66 pp.

Lang, A. *et al*. 1974. *Effects of Herbicides in South Vietnam. A. Summary and Conclusions*. Washington: National Academy of Sciences, [398] pp. + 8 maps.

Leighton, M. and Leighton, D. R. 1983. Vertebrate responses to fruiting seasonality within a Bornean rain forest. In: Sutton, S. L., Whitmore, T. C., & Chadwick, A. C. (eds). *Tropical Rain Forest: Ecology and Management*. Oxford: Blackwell Scientific Publications, 498 pp.: pp. 181–196.

Medway and Wells, D. R. 1976. *Birds of the Malay Peninsula. V. Conclusion, and Survey of Every Species*. London: H. F. & G. Witherby, 448 pp. + 24 pl. + 1 map.

Nowak, R. M. 1976. Wildlife of Indochina: tragedy or opportunity? *National Parks & Conservation Journal*, Washington, **50**(6): 13–18.

Orians, G. H. and Pfeiffer, E. W. 1970. Ecological effects of the war in Vietnam. *Science*, Washington, **168**: 544–554.

Raven, P. H. *et al*. 1980. *Research Priorities in Tropical Biology*. Washington: National Academy of Sciences, 116 pp.

Ricklefs, R. E. 1979. *Ecology*. 2nd ed. Concord, Mass.: Chiron Press, 966 pp.

Terborgh, J. and Winter, B. 1980. Some causes of extinction. In: Soulé, M. E. & Wilcox, B. A. (eds). *Conservation Biology: An Evolutionary-ecological Perspective*. Sunderland, Mass.: Sinauer Associates, 395 pp.: pp. 119–133.

Westing, A. H. 1976. In: SIPRI, *Ecological Consequences of the Second Indochina War*. Stockholm: Almqvist & Wiksell, 119 pp. + 8 pl.

Westing, A. H. 1982. Environmental aftermath of warfare in Viet Nam. In: *World Armaments and Disarmament, SIPRI Yearbook 1982*. London: Taylor & Francis, 518 pp.: pp. 363–389.

Westing, A. H. and Westing, C. E. 1981. Endangered species and habitats of Viet Nam. *Environmental Conservation*, Geneva, **8**: 59–62.

Whitmore, T. C. 1975. *Tropical Rain Forests of the Far East*. New York: Oxford University Press, 282 pp.

Chapter Four
Soil Ecology

4.A. Soil ecology: Symposium summary[1]

Hoàng Van Huây et al.[2]
University of Hanoi

This summary report examines the long-term effects on soil ecology (including the physical, chemical and biological properties of soil) of the herbicides applied to South Viet Nam during the Second Indochina War of 1961–1975.

The use of these chemicals caused heavy damage and long-term effects on soil ecosystems. In turn, these consequences could affect agricultural and forest production and, ultimately, human health. The effects of herbicides on soil may be direct or indirect. Their direct effects occur when they enter the soil and influence soil organic matter degradation processes or the soil microbiology. Their indirect effects occur through changing the vegetation and through their influence upon soil properties. The magnitude of changes in soil properties will vary depending upon other factors which influence the soil ecosystem, for example, the geological conditions, the topography, and the degree of development of the soil.

The Working Group heard and discussed presentations which dealt with the following three main soil topics: (*a*) the changes in soil properties that have occurred since the wartime spraying; (*b*) the effect of the herbicides on the ecosystem of soil micro-organisms; and (*c*) the fate of herbicides entering the soil, including the processes and products of their degradation.

The five major points made in these reports follow:

1. A large proportion of the elements of site fertility in undisturbed tropical forests is contained in the trees relative to the soil. Herbicides bring about a sudden return to the soil of the foliage of vegetation with its elemental content. Rapid decay of this detritus brings a flush of organic matter, nitrogen compounds and associated mineral elements to the soil. The increase in these materials changes the soil properties. The changes may be temporary or long-lasting depending upon many factors, such as rate of recovery of original vegetation, amount of conversion to other types of vegetation, nature of the geologic substrate, the topography and the degree of erosion.

[1] Summary report of a Working Group of the International Symposium on Herbicides and Defoliants in War: the Long-term Effects on Man and Nature, Ho Chi Minh City, 13–20 January 1983 (appendix 3).
[2] The Working Group consisted of: V. A. Bolshakov, Lê Van Can, Lê Trong Cúc (Vice-chairman), H. B. Gunner, Hoàng Van Huây (Chairman), C. F. Jordan, S. K. Mukherjee, V. V. Nazarov, Nguyên Hung Phuc, J. Pospisil, Pham Van Ty and P. J. Zinke (Rapporteur).

2. Loss of soil fertility elements may occur depending upon the intensity and duration of vegetational change induced by the herbicides. Repeated herbicide applications resulted in greater opening of the forest and conversion to other types of land occupance or use. The fertility content of the site (in both soil and vegetation) became less with the sequence from forest to grassland or bamboo. The soil fertility elements most susceptible to loss are potassium and nitrogen; a drop in available phosphorus also occurs owing to incorporation into insoluble substances.

3. A study made in the A Lưới valley in Binh Tri Thien province (in the former Thua Thien province, South Viet Nam, Military Region I) reported on changes in soils that had been collected from areas more than a decade after they had been converted from tropical inland forest to *Imperata* grassland through spraying (Huây & Cu, paper 4.B). The *Imperata* cover had been maintained during this period through the influence of periodic fires of anthropogenic origin. Where the topography was steep, the changes in soil properties included lower organic matter content and lower nitrogen content; also apparently less available phosphorus, and lower calcium, magnesium and iron in the soil's cation-exchange complex; they further may have included increases in acidity and aluminium content. On the other hand, where the topography was flat, as in the valley bottoms with alluvial soils, there were increases in the soil organic matter and nitrogen contents.

A study made of mangrove forest soils on the Ca Mau peninsula in Nam Can district of Minh Hai province (in the former An Xuyen province, South Viet Nam, Military Region IV) reported on long-term changes in areas which had been cleared of mangroves by the wartime spraying (Huây & Cu, paper 4.B). These soils were reported to have increased carbon and nitrogen contents, a greater acidity, less available phosphorus and lower potassium in comparison with uncleared mangrove forest. Where cleared soils were being used for agriculture, there was a drop in nitrogen content, but organic matter remained high. Soil deterioration has occurred in some of these areas owing to acid sulphate formation.

4. Herbicides can enter the soil directly or can be transmitted through sprayed plants via root exudates. In either case they may affect the species composition of the soil micro-organisms. There may be a selection in favour of those species which can decompose the compounds applied. These organisms will aid in the decomposition of the herbicides, but the possibility exists that the degradation compounds formed will be toxic.

5. With the appropriate combination of soil microflora present (usually the case), both 2,4-D and 2,4,5-T are fairly quickly degraded to non-toxic products. However, picloram is more stable in soil, being detectable for up to about three years. The arsenic from cacodylic acid (dimethylarsinic acid) may remain in soil in a fixed condition. Decomposition rates of herbicides in soil will vary depending upon the physical properties of the soil, its acidity, its microfloral composition, and its clay content (to which the chemicals could adhere). One study which has reported lengthy persistence is perhaps explicable on the basis of such fixation to the clay fraction of the soil.

Recommendations

The Working Group makes the following seven proposals for the benefit of scientists interested in long-term international co-operation:

1. Total ecosystem studies are needed to understand the role of herbicides (as of other xenobiotics) in biogeochemical cycling and their effects upon soil fertility.

2. A survey should be made of land use in areas subjected to wartime spraying and of the resulting sequence of vegetational change. Such a survey should include areas of intense land deterioration owing to erosion.

3. Techniques of restoration of soils deteriorated by adverse aspects of herbicide use and of subsequent land use should be developed and applied. Special attention is needed for acid sulphate soil reclamation.

4. Studies are needed on the persistence of herbicides in soils and on the processes of their degradation. The role of micro-organisms in the decomposition and degradation of herbicidal materials needs further study. Studies of effects on microfloral composition are needed, as are studies on selected indices of herbicide presence such as nitrogen fixers, cellulose decomposers, and mycorrhizal and similar microfloral/plant-root associations.

5. Studies on special soil topics related to herbicide use are needed, for example, the possible catalytic effect of clay minerals on the photo-oxidation or other degradation of herbicides and the effects of herbicides on processes of soil laterization.

6. Studies should be made on persistence of dioxin contaminants of herbicides in soil and their possible movement in the food chain to humans.

7. It is recommended that international agencies such as the United Nations Educational, Scientific and Cultural Organization (Unesco), United Nations Environment Programme (UNEP), United Nations Development Programme (UNDP), and Food and Agriculture Organization of the United Nations (FAO), and the international scientific community collaborate with Vietnamese scientists on studies of the problems associated with the wartime use of herbicides.[3]

---

[3] Additionally, the Working Group recognized, as did the Symposium participants in plenary session, that there are far broader aspects of herbicide use in a global context than the effects of herbicide use during wartime in Viet Nam on soil conditions. Its response of outrage to large-scale wartime use of herbicides for crop destruction and forest defoliation should not deny the benefits of their use to farmers and workers in the forest during times of peace.

4.B. Long-term changes in soil chemistry following herbicidal attack

Hoàng Van Huây and Nguyên Xuan Cu[1]

University of Hanoi

I. Introduction

Large areas of forest and agricultural lands in South Viet Nam were attacked with herbicides during the Second Indochina War, especially from 1966 to 1969. Long-term effects on the soil can result from the indirect influence of disturbing the vegetational cover (erosion, and so on) as well as from the direct action of the applied chemicals (poisoning of the microflora, and so on). In the present study changes in soil chemistry have been investigated for several divers inland and coastal forest habitats more than a decade following the herbicidal attacks.

II. Materials and methods

Forest soils were sampled at a number of inland and coastal locations, listed below. Sampling was carried out by standard techniques, most often to a depth of 20 or 30 cm. Sprayed and unsprayed comparison sites (usually two of each) were matched as closely as possible.

Soil (alluvium) in level mountain valleys was sampled in Binh Tri Thien province (in the former Thua Thien province); soil (brown soil overlying alluvium) on level plateaux was sampled in Tay Ninh province; soil (red ferralite or ultisol) on steep mountain slopes was sampled in Binh Tri Thien province (here overlying sandstone) and also in Gia Lai-Kontum province (in the former Binh Dinh province) and Phu Khanh province (in the former Phu Yen province) (in the latter two, overlying granite), and soil (saline muck) on a coastal delta was sampled in Minh Hai province (in the former An Xuyen province).

Standard chemical soil analyses were employed. Total organic matter was determined by Chiurin's method; total nitrogen by Kjeldahl's method; total phosphorus by a molybdate-blue colour comparison, and available phosphorus, that is, phosphorus pentoxide (P_2O_5), by Oniani's method. Exchangeable

---

[1] Adapted by the editor from the authors' Symposium presentation.

calcium (Ca^{2+}) was determined by an edetic acid (EDTA) volumetric method; exchangeable magnesium (Mg^{2+}) also by an EDTA volumetric method; mobile iron (Fe^{3+}) by a sulphosalicylic acid colour comparison, and mobile aluminium (Al^{3+}) by Xocolov's method. Hydrogen ion (H^+) content, that is, acidity, was determined in three ways: the so-called active hydrogen by electrometric measurement (in units of pH) of a potassium chloride (1 N KCl) extract; the so-called exchangeable hydrogen by a volumetric titration measurement of a potassium chloride extract; and the so-called hydrolytic hydrogen by a similar titration of a calcium acetate ($Ca[CH_3COO]_2$) extract.

III. Results

Inland forest

Total organic matter content of the soil was found to be higher in previously sprayed than unsprayed regions of inland forest in level valley sites that were surrounded by mountains (table 4.B.1), little affected in level plateau areas (table 4.B.2) and depleted on steep slopes (tables 4.B.3 and 4.B.4). These findings suggest that the initial disturbance of the plant cover resulted in substantial soil erosion and perhaps also in some soil creep. The subsequent replacement vegetation may also have been less immune to erosive forces. The results suggest that it would take some 26 t/ha of organic matter today in order to restore the top 20 cm of the soil on steep slopes to their prespray condition in this regard (table 4.B.3). As might be expected, total nitrogen content followed a pattern more or less similar to that of total organic matter content.

Table 4.B.1. Effect on soil (alluvium) of spraying herbicides on level forested mountain valleys

	Unsprayed[a]	Sprayed[b]	Change (per cent)	To equalize[c] (kg/ha)
Organic matter, total (g/kg)	20.6	49.9	+142	
Nitrogen, total (g/kg)	2.1	3.1	+48	
Phosphorus, total (g/kg)	0.43	0.40	−7	80
Phosphorus, available (mg/kg)	80.0	61.9	−23	48
Calcium, exchangeable (units/kg)	4.4	5.6	+27	
Magnesium, exchangeable (units/kg)	3.6	3.4	−6	
Iron, mobile (mg/kg)	796	255	−68	
Aluminium, mobile (mg/kg)	28.7	11.1	−61	
Acidity, active (pH)	4.4	5.2	+11	
Acidity, exchange (units/kg)	?	2.6	−	
Acidity, hydrolytic (units/kg)	?	13.8	−	

[a] One sampling location, place not specified: slope, 2 per cent; top 30 cm; cover, trees and grass.
[b] Average of two sampling locations in A Lưới district and Aso State Forest (both Binh Tri Thien province): slope for both, 2 per cent; top 10 cm and top 30 cm; cover for both, dead trees plus dense grasses.
[c] Amount required to increase the amount in the top 20 cm of the sprayed site to that in the unsprayed site, assuming a soil weight of 1 325 kg/m³.

Table 4.B.2. Effect on soil (brown soil overlying alluvium) of spraying herbicides on rather level forested plateaux

	Unsprayed[a]	Sprayed[b]	Change (per cent)	To equalize[c] (kg/ha)
Organic matter, total (g/kg)	38.4	39.5	+3	
Nitrogen, total (g/kg)	3.3	2.4	−27	2 400[d]
Phosphorus, total (g/kg)	0.20	0.45	+125	
Phosphorus, available (mg/kg)	82.5	56.9	−31	68
Calcium, exchangeable (units/kg)	2.5	2.9	+16	
Magnesium, exchangeable (units/kg)	1.5	3.4	+127	
Iron, mobile (mg/kg)	750	360	−52	
Aluminium, mobile (mg/kg)	5.2	5.6	+8	
Acidity, active (pH)	5.0	5.1	+1	
Acidity, exchange (units/kg)	0.8	1.4	+75	
Acidity, hydrolytic (units/kg)	6.0	7.0	+17	

[a] One sampling location in Tan Bien district (Tay Ninh province): slope, 3–5 per cent; top 30 cm; cover, trees.
[b] Average of two sampling locations, both in Tan Bien district: slope for both, presumably 3–5 per cent; top 15 cm for both; cover for both, dead trees plus dense *Imperata cylindrica* grass.
[c] Amount required to increase the amount in the top 20 cm of the sprayed site to that in the unsprayed site, assuming a soil weight of 1 325 kg/m³.
[d] This 2.4 t/ha of nitrogen would be 11.3 t/ha in the form of ammonium sulphate or 5.1 t/ha as urea.

Table 4.B.3. Effect on soil (red ferralite overlying sandstone) of spraying herbicides on steep forested mountain slopes

	Unsprayed[a]	Sprayed[b]	Change (per cent)	To equalize[c] (kg/ha)
Organic matter, total (g/kg)	30.8	20.8	−32	26 000
Nitrogen, total (g/kg)	3.4	2.0	−41	3 700[d]
Phosphorus, total (g/kg)	0.85	0.65	−24	530
Phosphorus, available (mg/kg)	34.2	55.6	+63	
Calcium, exchangeable (units/kg)	5.9	5.5	−7	
Magnesium, exchangeable (units/kg)	14.5	9.5	−34	
Iron, mobile (mg/kg)	1 020	650	−36	
Aluminium, mobile (mg/kg)	3.8	35.4	+832	
Acidity, active (pH)	4.9	4.9	0	
Acidity, exchange (units/kg)	1.5	1.5	0	
Acidity, hydrolytic (units/kg)	5.3	36.9	+596	

[a] Average of two sampling locations, both in A Lưới district (Binh Tri Thien province): slope for both, 27–36 per cent; top 20 cm and top 30 cm; cover for both, dense forest.
[b] Average of two sampling locations, both in A Lưới district: slope for both, 27–36 per cent; top 15 cm and top 20 cm; cover for both, dead trees plus grasses and herbaceous weeds.
[c] Amount required to increase the amount in the top 20 cm of the sprayed site to that in the unsprayed site, assuming a soil weight of 1 325 kg/m³.
[d] This 3.7 t/ha of nitrogen would be 17.5 t/ha in the form of ammonium sulphate or 7.9 t/ha as urea.

Table 4.B.4. Effect on soil (red ferralite overlying granite) of spraying herbicides on steep forested mountain slopes

	Unsprayed[a]	Sprayed[b]	Change (per cent)	To equalize[c] (kg/ha)
Organic matter, total (g/kg)	47.0	15.8	−66	83 000
Nitrogen, total (g/kg)	4.2	2.6	−38	4 200[d]
Phosphorus, total (g/kg)	1.25	0.80	−36	1 200
Phosphorus, available (mg/kg)	47.1	22.2	−53	66
Calcium, exchangeable (units/kg)	12.0	8.4	−30	
Magnesium, exchangeable (units/kg)	37.0	24.0	−35	
Iron, mobile (mg/kg)	?	1 160	−	
Aluminium, mobile (mg/kg)	?	0.2	−	
Acidity, active (pH)	?	5.0	−	
Acidity, exchange (units/kg)	?	1.3	−	
Acidity, hydrolytic (units/kg)	?	1.5	−	

[a] Average of two sampling locations, both in An Khe district (Gia Lai-Kontum province): slope for both, 18–36 per cent; top 20 cm for both; cover for both, dense forest.
[b] Average of two sampling locations, both in Deo Ca district (Phu Khanh province): slope for both, 27–36 per cent; top 10 cm and top 30 cm; cover for both, dead trees plus grass and shrubs.
[c] Amount required to increase the amount in the top 20 cm of the sprayed site to that in the unsprayed site, assuming a soil weight of 1 325 kg/m³.
[d] This 4.2 t/ha of nitrogen would be 19.8 t/ha in the form of ammonium sulphate or 9.0 t/ha as urea.

Phosphorus content in the soil, whether total or available, was generally found to vary only moderately and rather unpredictably from previously sprayed to unsprayed inland forest sites. The same must be said for exchangeable calcium and magnesium. No clear pattern emerged for mobile aluminium either, although here it should be noted that the data for red ferralite soil overlying sandstone on steep slopes suggest the possibility of a sharp increase as a result of spraying (table 4.B.3).

The mobile iron content of inland forest soils was found to be lower in the previously sprayed than unsprayed sites in three of the sites examined (tables 4.B.1, 4.B.2 and 4.B.3). This may have been the result of a change from reductive to oxidative conditions following the initial destruction of plant cover.

Soil acidity, at least in terms of the so-called active and readily exchangeable hydrogen ions, seems to have been little affected as a result of previous spraying of inland forest sites.

Coastal forest

The saline soil of mangrove forest sites was found to follow a pattern rather similar to that of inland forest sites (table 4.B.5). Organic matter content was virtually unaffected by previous spraying, in keeping with the level character of this terrain; mobile iron was found to be substantially reduced; acidity remained unaffected, and the other elements tested varied moderately in seemingly haphazard fashion.

Table 4.B.5. Effect on soil (saline muck) of spraying herbicides on level forested mangrove delta

	Unsprayed[a]	Sprayed[b]	Change (per cent)	To equalize[c] (kg/ha)
Organic matter, total (g/kg)	95.2	102.9	+8	
Nitrogen, total (g/kg)	2.0	5.0	+150	
Phosphorus, total (g/kg)	0.70	0.60	−14	260
Phosphorus, available (mg/kg)	22.5	103.0	+358	
Calcium, exchangeable (units/kg)	17.5	21.3	+22	
Magnesium, exchangeable (units/kg)	75.0	91.3	+22	
Iron, mobile (mg/kg)	2 040	280	−86	
Aluminium, mobile (mg/kg)	0.2	0.5	+150	
Acidity, active (pH)	6.0	5.8	−2	
Acidity, exchange (units/kg)	3.3	3.3	0	
Acidity, hydrolytic (units/kg)	3.5	3.8	+9	

[a] One sampling location in Nam Can district (Minh Hai province): slope < 1 per cent; top 40 cm; cover, mangrove forest.
[b] One sampling location in Nam Can district: slope < 1 per cent; top 40 cm; cover, dead trees.
[c] Amount required to increase the amount in the top 20 cm of the sprayed site to that in the unsprayed site, assuming a soil weight of 1 325 kg/m³.

IV. Conclusion

Examination of forest soils on several divers inland and coastal sites that had been massively sprayed with herbicides more than a decade previously suggests that substantial erosion, perhaps together with soil creep, had occurred on steep sites, and, conversely, that soil accumulation had occurred on valley sites in mountainous areas. This conclusion is suggested especially by the present pattern of organic matter content (low on slopes, high in valleys and normal in level areas). It was additionally found that of the several soil elements tested for, mobile iron stood out as appearing to have been reduced in amount on a long-term basis as a result of the prior spray attacks, as evidenced on all of the inland and coastal sites, whether steep or level.

4.C. Soil ecology: an overview

Paul J. Zinke
University of California

I. Introduction

Soil determines the productive capacity of the land. The crops and forests which provide the renewable resources for human populations are derived from the produce of plants grown in soil. Human actions that alter soil properties so as to decrease soil productivity are matters of universal concern.

There are a number of well-known adverse effects of warfare upon soil and its productivity (Westing, 1976, pages 64–68). One major aspect of deterioration is the broad-scale displacement of agricultural populations with a knowledge about the husbandry of local soils, who are in turn forced to subsist on new areas without traditional knowledge appropriate to the soil. The flow of refugees from war zones over-populating soils in adjacent areas brings about what could be referred to as the mining of the fertility of these soils through emergency cropping methods. These usually involve the shortening of traditional fallow periods in shifting slash-and-burn agriculture below the length of time needed to obtain adequate restoration of soil fertility. The direct effect of combatants in warfare, disturbing soils through shells, bombs and the cutting of forests to remove their advantages to the opposing side are further influences which destroy site fertility. Such destruction has occurred in warfare since time immemorial. Herbicides, however, have supplied a new increment to the wartime disturbance of soil and vegetation, the primary subject under consideration here.

Thus, the major concern here is with the disturbing effect of war upon soil which occurred in Viet Nam with the introduction of the wide-scale use of herbicides as forest defoliants and as crop-destroying agents. The objective is to evaluate the effects that the use of herbicides had upon soil and thus land productivity, and the longevity and possible deleterious translocations of toxic contaminants of the herbicides in soil and its vegetation. The effects of herbicides are *direct* when added to the soil, or *indirect* when the vegetation which that soil produces is removed along with its revitalizing influence on soil fertility.

II. Direct effects of herbicides on soil

The major direct effects of herbicides on soil are determined by their storage in soil in amounts which affect vegetative growth, the duration of this storage, the

75

nature of the decomposition products and the subsequent translocation of these materials from the soil to the flora which that soil supports, whether above-ground macroflora or sub-surface microflora. In addition, there is concern with contamination of waters infiltrating or running off that soil. Concerns arise from the herbicides as well as from any contaminants, all of which can be referred to as 'xenobiotics' (i.e., organic compounds foreign to the soil-vegetation system).

The amount of herbicides directly reaching the soil depends upon the density of the vegetative cover and its interceptive capacity. A large proportion of liquid materials falling onto a forest are trapped by a completely closed canopy of forest or other vegetational cover (Zinke, 1967). Few of the droplets reach the soil directly, except where openings occur in the canopy. Thus the amounts of herbicide falling through a closed forest canopy and directly affecting a soil will probably be less than 10 per cent of the application dose. However, where areas are barren of vegetation essentially the entire application will reach the soil. In areas where the herbicides were used for military defoliation or crop destruction, the application rates were of the order of 29 kg/ha (in terms of active ingredients). In the case of Agent Orange this amount was comprised of 13 kg/ha 2,4-D, and 16 kg/ha 2,4,5-T. These rates would have to be multiplied by the number of times a site was sprayed. For the 1.7 million hectares of inland forest area of Viet Nam that was sprayed, about 66 per cent was sprayed once, 22 per cent twice, 8 per cent three times, and 4 per cent four or more times (Westing, table 1.4; see also Lang et al., 1974, pages III-24 and IV-54, who provide very similar values on the basis of an about 85 per cent sample). The direct application to soils in these areas would be dependent upon amounts of throughfall, tree canopy interception, and subsequent additions from leaching and leaf fall.

III. Persistence of herbicides and their contaminants in soil

Herbicides

Papers in the literature concerned with possible deleterious effects of herbicides added to soil have been concerned with evaluating the persistence of herbicides as a potential inhibitor of crops on the site. The persistence of herbicides has been measured either by actual soil analysis for the herbicides or by a biological assay using crop plants common to an area. Persistence depends in part upon soil moisture and temperature as determined by local climate. Blackman et al. (1974) found at study sites similar to those in Viet Nam that most agricultural crops could be grown in soils treated with the military herbicides after six months of wet season, but that decomposition to insignificance was slower during the dry season.

Decomposition of the aromatic ring of the herbicides 2,4-D and 2,4,5-T apparently cannot take place in plants, but must take place in soil by the action of eubacteria that can destroy the phenol ring. Many soil organisms decompose 2,4-D or the chlorophenoxy acids added from fallen leaves of treated plants.

However, 2,4,5-T has a higher persistence in soil, as do other such trichloro-substituted phenoxy acids, apparently because of a need for commensal micro-organismic activity involving the presence of two or more species of micro-organisms. For example, neither *Pseudomonas* nor *Achromobacter* alone can decompose 2,4,5-T, but these bacteria can do so when present together. Since soil micro-organisms are responsible for completing cycles of major fertility elements on a site, herbicide effects may block processes in cycles responsible for the availability of these elements. Nitrogen transformations would be parti-cularly sensitive to this. The organophosphate insecticide diazinon, for example, results in an ecological succession of soil micro-organisms including the bacterium *Arthrobacter* (which decomposes carbon rings) and culminating in a predominance of the bacterium *Streptomyces* (Gunner, 1970). Each micro-organism in the series is responsible for the destruction of a different part of the diazinon molecule, and when only one is present, the diazinon is only in-completely destroyed.

The by-products of partial decomposition can be very toxic. Thus when a pesticide is added to the soil one must not only be concerned with the parent compound, but also with the breakdown products, which may be more toxic than the parent compound. The deleterious effect of the xenobiotic could be transferred to other trophic levels of organisms, for example, from *Arthrobacter* to protozoans to tubifex worms to fish. Moreover, if fungi are adversely affected by applied pesticides there may be changes in the binding capacities of poly-saccharides of the fungal filaments, to a point that seriously affects soil aggregation.

Nguyên Hung Phuc *et al.* (College of Medicine, Hanoi, unpublished) have recently found arsenic in the soil in areas treated by cacodylic acid (Agent Blue) roughly 15 years earlier. Although they pointed out that organic arsenic com-pounds may not be toxic as such, they may be transformed at some time into trivalent arsenic that is toxic. They also found chlorophenol degradation products of Agent Orange sprayed during that period still present in the soil.

Contaminants

A common contaminant of herbicides used in Viet Nam was dioxin, an impurity in the 2,4,5-T component of Agent Orange. The dioxin content of the Agent Orange used in Viet Nam is estimated to have averaged 4 g/m³, although occasional lots contained between 10 and 20 times this amount (Westing, chapter 1; Young *et al.*, 1978). Thus, the average amount of dioxin applied in a single spraying of Agent Orange was almost 110 mg/ha.

Dioxin has the highest persistence of any of the xenobiotics applied. For example, field studies in a heavily dioxin-contaminated spray site in Florida, USA have demonstrated that traces of dioxin can still be recovered from the top 15 cm of the soil 14 years after application (Young, 1983). It has been estimated that in the soil dioxin has a half-life of perhaps 3 to 5 years (Olie, paper 9.B; Westing, 1982, page 368).

Aquatic life, such as fish and shrimp, were found by Baughman & Meselson (1973) to contain about 100 ng/kg of dioxin on a dry weight basis in inland regions near areas where spraying had occurred. This raised the matter of translocation from the soil to living organisms. Plants grown in nutrient culture having 160 μg/kg dioxin were found to contain up to 1.5 mg/kg on a dry weight basis (Isensee & Jones, 1971). Oats and soya beans grown on sandy loam soils containing dioxin also contained some dioxin, although little translocation took place from the older to younger plant parts (Isensee & Jones, 1971).

Soil samples recently obtained from inland soils and sediments in mangrove areas sprayed about a decade earlier were found to contain up to about 30 ng/kg of dioxin (Olie, paper 9.B). This has led to concern over the long-term contamination of fish, a matter being investigated by Bui Thi Lang (Committee for Science and Technique, Ho Chi Minh City).

IV. Indirect effects of herbicides on soils

The greatest concern with regard to herbicidal effects on soils has been the indirect effects brought about by disturbing the normal cycling of elements between vegetation and soil. This is particularly appropriate in forest areas where the maintenance of soil fertility is largely a function of the process of return of fertility elements to the soil through leaf fall and foliar leaching.

The indirect effect of herbicides on soil fertility will depend upon the degree to which defoliation and mortality have occurred owing to herbicides. This in turn is related to the number of times that herbicides have been used on a single area. Earlier it was noted that about 34 per cent of the areas sprayed in upland forests had been sprayed twice or more.

A considerable proportion of the fertility of a closed forest ecosytem is contained in the vegetation (table 4.C.1). The immediate effect of herbicides in inducing leaf fall would be to drop the material stored in the foliage to the ground where decomposition would result in its return to the soil. A relatively high proportion of the nitrogen and potassium of the site, as well as of its phosphorus, is stored in the vegetatior. The extent to which these elements are retained in the soil against surface water run-off, erosion or leaching is critical. Rapid vegetative regrowth on the site lessens the chance for fertility loss. Moreover, there is less fertility storage in secondary successional vegetation such as bamboo than in the original vegetation. A bamboo stand measured near Mae Moh, Thailand had 0.8 kg/ha of nitrogen and 0.5 kg/ha of potassium in the above-ground vegetation (Zinke, 1974).

A major use of upland forests in mountainous areas of Viet Nam is shifting slash-and-burn cultivation. During the period of herbicide use there was a loss of forest fallow cover crop which is essential to maintaining this form of agriculture. This effect was analogous to shortening the fallow cycle, which tends to reduce soil fertility (Zinke et al., 1978).

Huây & Cu (paper 4.B) recently conducted soil studies in the A Luói valley, Binh Tri Thien province, an area near the Laotian border west of Danang which

Table 4.C.1. Comparative storage of selected fertility elements in the tropical inland closed-forest ecosystem at two sites in Thailand

	At Ban Pa Prae		At Sakerat	
	(kg/ha)	(per cent)	(kg/ha)	(per cent)
A. Nitrogen				
Vegetation	13.7	14	5.3	6
Foliage	6.2	6	1.0	1
Wood	7.5	8	4.3	5
Soil (to 1 m)	86.7	86	86.7	94
Total ecosystem	**100.4**	**100**	**92.0**	**100**
B. Potassium				
Vegetation	15.4	65	6.8	45
Foliage	5.0	21	0.8	5
Wood	10.4	44	6.0	40
Soil (to 1 m)	8.2	35	8.2	55
Total ecosystem	**23.6**	**100**	**15.0**	**100**

Source: Zinke (1974, page 13).

formerly was densely forested and had received repeated herbicide applications during the Second Indochina War. In comparing sprayed with unsprayed sites, they found that the organic matter and nitrogen contents of the soils on steep slopes were depressed more than a decade after the applications, but were elevated in the soils of mountain valleys. This was concluded to be the result of erosion from the slopes and deposition in the alluvial areas. In addition, soil at all the sites exhibited lowered mobile iron. Forest vegetation had been reduced in density on the sprayed sites or else had been replaced by bamboos or grasses such as *Imperata* and *Saccharum*, further indicating a decreased fertility.

Similar effects were reported by Huây & Cu (paper 4.B) in a study on the soils of a mangrove forest which had been destroyed by herbicides on the Ca Mau peninsula, Minh Hai province at the southern end of Viet Nam. In another study of this mangrove forest, Phan Nguyên Hong *et al.* (Teachers Training College, Hanoi, unpublished) reported increases in organic matter in the soils of herbicide-destroyed areas, particularly in depressions, and a tendency towards more acid soil conditions, the pH declining from 7.2 to 6.8. In rear (back) mangrove areas of *Melaleuca* this resulted in a tendency towards acid sulphate soil formation. Where the mangrove forest had been killed it was also noted that calcium, magnesium and potassium contents of the soil dropped while sodium, sulphate and iron contents became higher.

V. Field observations

The author of this paper travelled in January 1983 to areas typical of defoliated inland forest and mangrove forest which he had previously visited and studied in 1970–1971 as a member of the US National Academy of Sciences Committee on

Herbicides in Vietnam. In the inland forests, where soils were productive and defoliation had created the most openings, new settlers have since frequently taken advantage of the land, and were found to be growing crops of sesame and rice in lowland forest areas on alluvial soils, and cassava on upland areas. An area of previous shifting slash-and-burn cultivation was observed on the road from Ho Chi Minh City to Dalat in north-eastern Dong Nai province (in the former Long Khanh province) that had been sprayed with herbicides at the fallow forest stage. This forest has now regrown, but with much bamboo.

The trip to a mangrove site began in the north-eastern part of the coastal Rung Sat region, south-east of Ho Chi Minh City at the northern end of Go Guy River, and went about 6 km south-westerly through a large area where multiple sprayings with Agent Orange had occurred in 1967, 1968, 1969 and 1970 (Lang *et al.*, 1974, page IV-99). In this area, the forester currently in charge (January 1983) had organized planting crews of several hundred people, each person planting 0.1–0.2 ha during a 3-h period between high tides. Soil preparation consisted of clearing stumps and trash for firewood. Sprouted *Rhizophora* seeds from the Ca Mau peninsula were used. Plantings are done at densities of 4 000/ha, 6 000/ha or 8 000/ha. Within a few years the interlocking roots protect the soil from erosion. In three districts totalling 15 000 ha, 20 per cent of the area had been planted. Trees that had been planted three years earlier were more than 2 m tall.

The general conclusion from the inland and coastal field trips was that recovery is taking place, either through intensive forest management efforts or through the agricultural efforts of individual farmers where the soil was still productive. Areas of non-productive soil, whether naturally that way or owing to herbicide treatment, have remained in poor stands of grass cover or bamboo.

VI. Questions remaining

Many questions remain concerning the effects of herbicides on soil fertility in the tropics, and the effects of the repeated use of them during the Second Indochina War. It is essential that these be answered because of the utility of herbicide use in tropical agriculture and forestry. To deny such use because of lack of answers, or because of the emotions attached to the wartime use of herbicides, would be unfortunate.

A survey needs to be made of land use in herbicide-treated areas and the resulting sequences of vegetational and soil change that have occurred. The survey should locate any areas of intensive land deterioration owing to erosion or to other processes of fertility loss.

Techniques of restoration need to be developed and applied to soils which have been identified as having deteriorated owing to herbicide use and any resulting land use changes. In this regard, special attention is needed for the reclamation of acid sulphate soils.

Studies are needed of the persistence of herbicides in soils. The role of microorganisms in the decomposition of these compounds needs to be studied. An elucidation is needed of the effects of microfloral composition. Selected indices of herbicide presence are needed such as effects on nitrogen-fixing organisms, cellulose decomposers and mycorrhizal (fungus/root) associates.

Studies on topics especially related to herbicide chemistry in the soil are needed. For example, an understanding is needed of the role of the catalytic effect of clay minerals on photo-oxidation and degradation of the compounds, and of the effects of herbicides upon such soil-evolving processes as laterization and podzolization.

Highly toxic materials such as dioxin need special attention in research work aimed at understanding the processes of degradation, uptake by plants and movement in food chains which affect humans.

The indirect effects of herbicides require studies of the entire soil–hydrosphere–vegetation system. Exactly what happens to defoliated and dead vegetation requires elucidation in the context of geochemical cycling and the resultant effects on soil fertility. These effects need to be established for the varying conditions that occur owing to the different soil–vegetation associations present.

References

Baughman, R. and Meselson, M. 1973. Analytical method for detecting TCDD (dioxin): levels of TCDD in samples from Vietnam. *Environmental Health Perspectives*, Research Triangle Park, N. Car., **1973**(5): 27–35.

Blackman, G. E., Fryer, J. D., Lang, A. and Newton, M. 1974. *Effects of Herbicides in South Vietnam. B[3]. Persistence and Disappearance of Herbicides in Tropical Soils.* Washington: National Academy of Sciences, 59 pp.

Gunner, H. B. 1970. Microbial ecosystem stress induced by an organophosphate insecticide. *Mededelingen Faculteit Landbouw Wetenschappen*, Ghent, **35**(2): 581–597.

Isensee, A. R. and Jones, G. E. 1971. Absorption and translocation of root and foliage applied 2,4-dichlorophenol, 2,7-dichlorodibenzo-*p*-dioxin, and 2,3,7,8-tetrachlorodibenzo-*p*-dioxin. *Journal of Agricultural and Food Chemistry*, Washington, **19**: 1210–1214.

Lang, A. *et al.* 1974. *Effects of Herbicides in South Vietnam. A. Summary and Conclusions.* Washington: National Academy of Sciences, [398] pp. + 8 maps.

Westing, A. H. 1976. In: SIPRI, *Ecological Consequences of the Second Indochina War.* Stockholm: Almqvist & Wiksell, 119 pp. + 8 pl.

Westing, A. H. 1982. Environmental aftermath of warfare in Viet Nam. In: *World Armaments and Disarmament, SIPRI Yearbook 1982.* London: Taylor & Francis, 518 pp.: pp. 363–389.

Young, A. L. 1983. Long-term studies on the persistence and movement of TCDD in a natural ecosystem. In: Tucker, R. E., Young, A. L. & Gray, A. P. (eds). *Human and Environmental Risks of Chlorinated Dioxins and Related Compounds.* New York: Plenum Press, 823 pp.: pp. 173–190.

Young, A. L., Calcagni, J. A., Thalken, C. E. and Tremblay, J. W. 1978. *Toxicology, Environmental Fate, and Human Risk of Herbicide Orange and Its Associated Dioxin.* San Antonio, Tex.: US Air Force Occupational & Environmental Health Laboratory Rept No. OEHL TR-78-92, 247 pp.

Zinke, P. J. 1967. Forest interception studies in the United States. In: Sopper, W. E. & Lull, H. W. (eds). *Forest Hydrology.* Oxford: Pergamon Press, 813 pp.: pp. 137–161.

Zinke, P. J. 1974. *Effects of Herbicides in South Vietnam. B[19]. Effect of Herbicides on Soils of South Vietnam.* Washington: National Academy of Sciences, 39 pp.

Zinke, P. J., Sabhasri, S. and Kunstadter, P. 1978. Soil fertility aspects of the lua' forest fallow system of shifting cultivation. In: Kunstadter, P., Chapman, E. C. & Sabhasri, S. (eds). *Farmers in the Forest: Economic Development and Marginal Agriculture in Northern Thailand.* Honolulu: University Press of Hawaii, 402 pp.: pp. 134–159.

Chapter Five
Coastal, Aquatic and Marine Ecology

D

5.A. Coastal, aquatic and marine ecology: Symposium summary[1]

Mai Dình Yên *et al.*[2]
University of Hanoi

This summary report examines the long-term effects on coastal mangrove, inland (freshwater) aquatic, and marine ecology of the herbicides applied to South Viet Nam during the Second Indochina War of 1961–1975.

Mangrove forests were frequently defoliated during the years between 1962 and 1971 and this resulted in the complete destruction of a significant percentage of these coastal forests from the Rung Sat area south-east of Ho Chi Minh City to the Ca Mau peninsula at the southern tip of the country (i.e., much of the eastern fringe of the former Military Region IV). A US National Academy of Sciences study team concluded in 1974 that the affected mangrove areas were so intensively damaged that natural recovery might take as long as 100 years, owing in part to the extensive loss of seed sources.

The destruction of the coastal mangrove forests resulted in a significant potential loss of timber, firewood, wood for charcoal, tannin, and other forest products. It also presumably led to a decrease in estuarine and near-shore fishery yields. In essence, a substantial proportion of the mangrove ecosystem, including its associated estuarine flora and fauna, experienced a significant productive loss.

The use of herbicides in the inland (upland) areas of South Viet Nam was more extensive than in the coastal mangrove areas, but the subsequent damage was more variable in the inland areas than in the coastal areas. Disturbance to the inland watersheds (catchments) and the introduction of herbicides into inland aquatic ecosystems have been associated with aquatic biological changes that are considered to be serious.

The Working Group reviewed the available information and data and recommends that countries and international organizations supporting the development of Viet Nam provide assistance in the following two areas: (*a*) for the

---

[1] Summary report of a Working Group of the International Symposium on Herbicides and Defoliants in War: the Long-term Effects on Man and Nature, Ho Chi Minh City, 13–20 January 1983 (appendix 3).
[2] The Working Group consisted of: Vo Trí Chung, D. F. Dunn, Phan Nguyên Hông, Bui Thi Lang (Vice-chairwoman), Nguyên Duy Minh, S. C. Snedaker (Rapporteur), S. Sokolova, Nguyên Van Tuyên and Mai Dình Yên (Chairman).

85

assessment and monitoring of any possible chronic effects from residual herbicides during the process of ecosystem recovery; and (b) for the evaluation of productive alternatives of economic and social benefit for the utilization of the altered habitats.

Coastal mangrove ecosystems

The data and information available for the mangrove ecosystem indicate that the effects of spraying are widespread, long-lasting, and severe within the affected areas (Hiêp, paper 5.B). The relatively good quality of the available information makes it suitable for the definition of the basic research programmes required to assist in the restoration of the habitat, including its flora and fauna, and for the directions that should be taken in developing new economic opportunities. Trial plantations of a high-value species of mangrove (*Rhizophora apiculata*) in its former habitat could accelerate the recovery of the mangrove ecosystem. However, degraded and otherwise unsuitable habitats will require the evaluation and selection of alternative economic uses.

Whereas it is doubtful that toxic residues of the herbicides persist in significant concentrations, there is a reasonable probability that the defoliated and altered inland watersheds continue to have an impact on the downstream coastal mangrove forests. Altered hydro-period, excessive erosion and deposition, and introductions of deleterious materials could have a significant effect on the fauna and flora of the mangrove ecosystem and associated estuarine areas. Insufficient quantitative data exist as yet to assist in evaluating these possible impacts.

Inland aquatic ecosystems

Compared to the existing knowledge of the mangrove ecosystem, the potential effects on the inland aquatic ecosystems have been much less well described in the literature. However, some relevant data and information have been assembled by qualified Vietnamese scientists (Yên & Quýnh, paper 5.C). These suggest the existence of spray-induced adverse effects, including the loss of freshwater vertebrate and invertebrate species and anomalous deformations among species of the local freshwater algae. Because many questions remain to be answered concerning this topic, a statistically valid assessment study is warranted that would determine the characteristics of the altered freshwater habitats and their biotic components, particularly those that have economic importance.

Marine ecosystems

Vietnamese scientists have confirmed earlier reports of declining marine fishery stocks and of the disappearance of certain species. Although similar problems are being reported in other countries of the region, the Vietnamese situation cannot be attributed to overfishing and related exploitative fishing practices. It is therefore urgent that fishery stock assessments be undertaken and that local training be provided in fishery management and capture techniques.

Recommendations

Owing to the ecological and economic values of Viet Nam's coastal, inland aquatic and marine ecosystems, and because Viet Nam's opportunities for natural resource development are limited, the Working Group makes the following six recommendations:

1. It is suggested that Viet Nam participate in the Regional Coastal and Marine Programme of the United Nations Educational, Scientific and Cultural Organization (Unesco). To this end, Viet Nam must create a co-ordinating National Mangrove Committee and should send members of this Committee to the Unesco Programme in Bangkok for training in mangrove biology and management. The National Committee should also serve as an *ad hoc* advisory body to monitor the reclamation and restoration of the altered ecosystems.

2. Viet Nam should solicit co-operation from the Czechoslovak Academy of Sciences concerning the use of insect indicator species in monitoring the recovery of inland aquatic ecosystems.

3. Vietnamese scientists and natural resource managers should actively solicit library materials, methodological handbooks, and training aids on relevant scientific and management studies.

4. Viet Nam should undertake statistically controlled studies of each altered ecosystem for the purpose of explaining why certain ecosystems appear to be slow in recovering, doing so in order to lay a scientific basis for accelerating the recovery processes.

5. Viet Nam should evaluate all alternative potential uses of altered ecosystems, with emphases on aquaculture and on the harvesting of species not previously utilized in the country.

6. Viet Nam should incorporate socio-economic considerations into its natural resource development plans in order to ensure that maximum human benefits are achieved.

In closing, the Working Group considers international co-operation with Vietnamese scientists a necessity in overcoming the adverse effects of the war.

5.B. Long-term changes in the mangrove habitat following herbicidal attack

Dinh Hiêp[1]

Institute of Forest Inventory and Planning, Hanoi

I. Introduction

Large areas of forest and agricultural lands in South Viet Nam were attacked with herbicides during the Second Indochina War, especially from 1966 to 1969. The coastal mangrove forests were no exception (Westing, 1982, page 373). Singled out for examination here is the mangrove forest of southernmost Viet Nam, considered the largest and perhaps most productive such habitat in the world.

II. Materials and methods

The study area was the almost 65 000-ha Ca Mau peninsula, which falls within Nam Can district, Minh Hai province (in the former An Xuyen province). The basic approach was to compare the situation just prior to 1965 (i.e., prior to most of the herbicidal attacks) with that in 1973 (several or so years after the attacks). This was done by analysing: (*a*) US Army Pictomap No. L8020, 1:25 000, 1965; and (*b*) US National Aeronautics and Space Administration Landsat satellite image No. N.1164-02510, 3 January 1973. Examination of these documents was facilitated by magnifying lenses, stereoscopes, and other photogrammetric devices. In each case, the land-type features observed were converted for comparison to maps having a scale of 1:100 000. A number of on-site inspections were carried out between 1975 and 1980, primarily in order to assist in interpreting the features obtained from the aerial photography and imagery. The following land types were distinguished: (*a*) mangrove forest, in turn separated into (*i*) *Rhizophora* and (*ii*) other mangroves; (*b*) treeless land; and (*c*) cultivated land.

---

[1] Adapted by the editor from the author's Symposium presentation. See also his paper 2.B.

III. Results

It was found that prior to 1965 some 53 000 ha of the Ca Mau peninsula were covered by *Rhizophora* forest, that is, 82 per cent of the peninsula (table 5.B.1). The wartime herbicidal destruction extensively reduced the area of this rich habitat, there being only about 20 000 ha in the year 1973. This 20 000 ha of extant *Rhizophora* includes a small (though here not determined) area of replanting, an operation that has been carried out in the region since 1971. The actual areal extent of military habitat destruction on the peninsula—that is, tree loss and site degradation—may thus have been more than 33 000 ha, that is, in excess of 50 per cent.

Table 5.B.1. Land-type distribution for Ca Mau peninsula, Nam Can district, Minh Hai province, pre-1965 and 1973

Land type	Pre-1965 (ha)	1973 (ha)	Change (per cent)
Mangrove forest	57 850	26 370	−54
Rhizophora	53 330	20 140	−62
Other	4 520	6 230	+38
Treeless land	7 050	36 440	+417
Cultivated land	0	2 090	+∞
Total	**64 900**	**64 900**	**0**

Of the 33 000 ha of land formerly occupied by *Rhizophora*, it appears that only about 1 000–2 000 ha had in 1973 been naturally repopulated by such pioneer species as the euphorbia *Excoecaria*; and an area of a similar size had been placed under cultivation. However, the remaining about 29 000 ha of former *Rhizophora* lands were bare of trees in 1973 (only a minute, here undetermined, fraction of this having been the result of subsequent felling). Field examinations as recent as 1980 have reinforced the observation that natural regeneration of *Rhizophora* has been minimal.

IV. Conclusion

The conclusion is that artificial replanting will be necessary in order to restore the destroyed *Rhizophora* forest within a reasonable time frame.

Reference

Westing, A. H. 1982. The environmental aftermath of warfare in Viet Nam. In: *World Armaments and Disarmament, SIPRI Yearbook 1982*. London: Taylor & Francis, 518 pp.: pp. 363–389.

5.C. Long-term changes in the freshwater fish fauna following herbicidal attack

Mai Dình Yên and Nguyên Xuân Quýnh[1]

University of Hanoi

I. Introduction

Large rural areas of South Viet Nam were attacked with herbicides during the Second Indochina War, especially from 1966 to 1969. Long-term effects on the indigenous freshwater (stream and lake) fish fauna can result from the indirect influence of a disrupted vegetational cover on the adjacent land areas, from herbicidal effects on the aquatic plant community (both macro and micro), from toxic effects on food species (zooplankton, zoobenthos, macro-invertebrates) and from direct toxic action to the fish themselves of the applied chemicals. In the present study, changes in the stream and pond fish fauna have been investigated for an inland (upland) forest habitat more than a decade following the herbicidal attacks. Elsewhere it has been reported that the freshwater tarpon *Megalops cyprinoides* was adversely affected in the Mekong Delta region by the wartime spraying (Westing, 1976, page 72).

II. Materials and methods

The previously sprayed area was the A Luói valley, a large valley primarily in A Luói district (and extending somewhat into Huong Hóa district), Binh Tri Thien province (in the former Thua Thien province). The valley is on the eastern slopes of the Truong Son mountain range, bordering Laos and roughly 50 km south of Quang Tri City and 100 km west of Danang. The area had been 80–90 per cent forested prior to the attacks, a substantial fraction of which was reduced to and remains grassland and other low herbaceous or woody vegetation.

The fish were recorded on two occasions, in 1981 and 1982, using seines and lift nets and also via purchases from local fishermen at the A Luói village market. The collections were made at two sites. The first source was the A Sáp River, a major stream in the A Luói valley that flows into the Sekong River (which in turn flows into the Mekong River). The A Sáp River is 50–70 m wide and meanders through the valley. It has a hard (pebbly) bottom except at the

---

[1] Adapted by the editor from the authors' Symposium presentation.

bends, where it is deep and muddy. It has a low-water period from November to May and a high-water period from June to October. The second source was a 2-ha pond (with a maximum depth of 6–7 m) formed by the Tabat dam across a small tributary of the A Sáp River, an impoundment made for purposes of irrigation during the dry season.

III. Results

A total of 24 species of fish were recorded (table 5.C.1), substantially less than would be found in a comparable unsprayed location. Indeed, even of the two dozen species identified, some half dozen were recognized as post-war colonizers. The catfish *Clarias fuscus* and loach *Misgurnus fossilis anguillicaudatus*, both from northern Viet Nam, seem to have been especially successful in colonizing the newly depauperate aquatic habitat.

Not only was species composition reduced and altered, but the total biomass of the fish fauna appears to have declined as well. This latter finding of diminished habitat productivity is based on reports from the local inhabitants. In the early

Table 5.C.1. Wild fishes of A Luói district, Binh Tri Thien province, 1981–1982 (partial listing)

Scientific name[a]	Family name[a]	Comment
Barilius pulchellus	Cyprinidae (carp)	Increased numbers
Barilius sp.	Cyprinidae (carp)	Increased numbers
Clarias fuscus	Clariidae (catfish)	Colonizer[c]
Ctenogobius ocellatus	Gobiidae (gobies)	
Danio regina	Cyprinidae (carp)	Colonizer[d]
Glossogobius giurus	Gobiidae (gobies)	
Hampala macrolepidota	Cyprinidae (carp)	Increased numbers
Mastacembelus armatus favus	Mastacembelidae (spiny eels)	Increased numbers
Mastacembelus sp.	Mastacembelidae (spiny eels)	Increased numbers
Misgurnus fossilis anguillicaudatus	Cobitidae (loaches)	Colonizer[c]
Mystus nemurus	Bagridae (catfish)	Increased numbers
Nemacheilus spiloptera	Cobitidae (loaches)	Colonizer[d]
Ophiocephalus gachua[b]	Ophiocephalidae (snakeheads)	
Ophiocephalus marulius[b]	Ophiocephalidae (snakeheads)	
Oryzias latipes	Cyprinodontidae (killifish)	
Osteochilus hasselti	Cyprinidae (carp)	Colonizer[c]
Puntius partipentazone	Cyprinidae (carp)	
Puntius semifasciolatus	Cyprinidae (carp)	
Puntius spp. (3)	Cyprinidae (carp)	
Rasbora lateristriata	Cyprinidae (carp)	
Rasbora trilineata	Cyprinidae (carp)	
Scaphiodonichthys sp.	Cyprinidae (carp)	

[a] Nomenclature is according to Orsi (1974), as far as possible.
[b] This may be the genus *Channa* (family Channidae) of Orsi (1974, page 164).
[c] Energetic post-war colonizer from northern Viet Nam.
[d] Post-war colonizer from the upper tributaries of the Mekong River. Certain (unspecified) species of *Puntius* have done this also.

1960s, prior to the herbicidal attacks against the A Lưới valley, one fisherman would routinely net 0.5–1.0 kg of fish from the A Sáp River in half a day, whereas now he obtains only 0.2–0.3 kg in this fashion.

The reduced fish biomass, and possibly the reduced number of fish species as well, can be attributed at least in part to reductions in their food supplies. The aquatic plants expected at the stream bends and in the pond were for the most part absent; and with them, many of the plant-feeding (macrophyto-phagous) fish. The aquatic invertebrate fauna—including the molluscs, crus-taceans and rotifers—were found to be reduced in numbers or actually missing. Thus, the indigenous species of shrimp (suborder Macrura), crabs (suborder Brachyura), snails (class Gastropoda), mussels (superclass Bivalvia) and water fleas (order Cladocera) were low in numbers. Moreover, various typical species of worms (class Oligochaeta), larval two-winged flies (family Chironomidae; suborder Nematocera) and mountain crabs (family Potamidae; suborder Brachyura) could not be found at all despite the systematic use of nets and bot-tom grabs. On the other hand, populations of various insects possessing aquatic larvae have maintained their numbers, among them the mayflies (order Plectoptera), stoneflies (order Plecoptera), caddisflies (order Trichoptera), and dragonflies (order Odonata). These in turn are assumed to account at least in part for the increased numbers of certain local fishes (table 5.C.1).

IV. Conclusion

The long-term effect on the local freshwater fish fauna brought about by the herbicidal decimation of more than a decade ago has taken essentially three forms: (*a*) a local reduction in fish species diversity; (*b*) invasion by a number of fish species previously alien to the sprayed area; and (*c*) a local reduction in fish biomass and productivity. These effects appear, at least in large part, to be the result of reductions in the fish's natural food supplies, both plant and animal.

References

Orsi, J. J. 1974. Check list of the marine and freshwater fishes of Vietnam. *Publications of the Seto Marine Biological Laboratory*, Kyoto, **21**: 153–177.
Westing, A. H. 1976. In: SIPRI, *Ecological Consequences of the Second Indochina War*. Stockholm: Almqvist & Wiksell, 119 pp. + 8 pl.

5.D. Coastal, marine and aquatic ecology: an overview

Samuel C. Snedaker

University of Miami

I. Introduction

Large areas of upland and coastal forests in South Viet Nam were defoliated or otherwise destroyed during the Second Indochina War in order to deny concealment to opposing military forces, and for various other military purposes (Buckingham, 1982). In general, inland (upland) forests exhibited variable effects in response to the use of chemical defoliants, whereas the coastal forests dominated by several species of mangrove were uniformly eliminated in those areas that were sprayed. The apparently high sensitivity of mangroves to herbicides, compared to non-halophytic tropical tree species, has been generally acknowledged (Truman, 1961–1962; Walsh *et al.*, 1973).

Although the primary objective of the wartime spray programme was the removal of dense forest cover, the aerial spray also fell in part into contiguous aquatic ecosystems such as streams and lakes in the upland areas and tidal channels in the coastal mangrove/estuarine areas. For these aquatic and semi-aquatic ecosystems, the full range of potential acute effects are matters of speculation, as are the longer-term chronic effects, including the important question of the rate at which the affected ecosystems will recover. Irrespective of these considerations, there is a continuing and widespread interest in the long-term chronic effects owing in part to the stability and environmental persistence of the dioxin contaminant of the herbicide 2,4,5-T in Agent Orange. In addition, there are lingering questions concerning the indirect chronic effects resulting from loss of littoral and intertidal vegetation in relation to detritus production and the role of the latter in fishery production.

In this report, there is a disproportionate emphasis on coastal ecosystems (i.e., mangrove forests) and marine ecosystems. This emphasis reflects both the magnitude of the impact of spray operations on the coastal and near-shore environments and the collective interests of Vietnamese and other scientists. In contrast, little is known about the basic ecology of Viet Nam's aquatic freshwater ecosystems and the effects on them of the herbicide spray operations. Paucity of comment therefore does not indicate the relative absence of environmental problems, but rather the relative paucity of information and data that can be interpreted in a rigorous manner.

II. Coastal ecosystems

Of all the habitat types in South Viet Nam, the coastal mangrove forests were considered to be the most heavily devastated by the wartime herbicide operations. Efforts to assess the areal extent of damage have revealed various estimates of the actual total pre-war area of mangroves in Viet Nam. This is due in part to the difficulty of defining a boundary between the true mangrove vegetation and the contiguous inland vegetation such as the *Melaleuca* forests or so-called rear (back) mangrove; and also to the problem of how much to include (if any) of low-density, scrubby mangrove areas. It appears, however, that a reasonable estimate for the total extent of the true mangrove habitat in South Viet Nam is 300 000 ha, or perhaps a little less (Maurand, 1943; Ross, 1974, page 30; Tung, 1967, page 30; Westing, 1976, page 5). More than 60 per cent is on or contiguous with the Ca Mau peninsula, which forms the southern tip of Viet Nam; and almost 20 per cent is in the coastal Rung Sat region south-east of Ho Chi Minh City (the Saigon River delta). The area of true mangroves that was sprayed with herbicides at least once is more or less consistently reported as approximately 124 000 ha (Westing, 1976, page 30), primarily in the Ca Mau and Rung Sat areas. Based on the fact that mangroves are extremely sensitive to herbicidal defoliants, this estimate suggests that more than 40 per cent of the total mangrove forest area experienced significant mortality during the Second Indochina War. Of course, when the *Melaleuca* forests and fringe areas of sparse mangroves are included, the total affected coastal area is significantly larger. For example, Westing (1976, page 30) reports that 27 000 ha of the 200 000 ha area in rear mangroves was also sprayed, inducing a high mortality of the vegetational dominants. These damage estimates do not, of course, include aquatic, estuarine or marine water areas subjected to spray or spray drift as part of the routine spray operations. There are essentially no data available upon which estimates can be developed for determining the impact on these water areas.

It appears that the pre-war mangrove forests of South Viet Nam were similar to other mangrove forests in South-east Asia in species composition, general forest structure, ecosystem functioning, and overall utilization (Cruz, 1979). The most important genera are *Avicennia*, *Bruguiera*, *Ceriops*, *Rhizophora*, and *Sonneratia*. The Ca Mau mangrove habitat was ecologically the most highly developed in Viet Nam and the trees were managed for commercial benefit using established silvicultural practices.

In the other major mangrove areas, particularly in the Rung Sat, the forests were less well developed, although they were heavily utilized for a variety of domestic purposes. Ross (1974) reports that 51 per cent of the Rung Sat was dominated by trees, whereas the remaining 49 per cent was distributed among areas defined as: cultivated; brush and herbaceous vegetation; bare areas (exposed soil, tidal mud flats); and surface water areas. The area of surface water was estimated at 23 per cent of the overall Rung Sat area. Ross held the opinion that the Rung Sat forests had ". . . been subjected to intense and often abusive exploitation", a phenomenon which is common within mangrove forests that are located close to areas having a high human population density. In all mangrove forest areas in Viet Nam, the most important mangrove species is *Rhizophora apiculata*, which has a variety of domestic and commercial uses owing mainly to the large size of the mature tree and the mechanical properties of its wood. In general, the mangrove forests in Viet Nam were, and continue to be, utilized for timber, fuel wood and charcoal, small dimension wood, and tannin-based dyes—albeit on a relatively limited basis at the present time.

Unlike inland forest areas, where damage was patchy, variable and difficult to assess, there is no question of the cause/effect relationship between herbicides and mangrove mortality. In general, defoliation, followed by mortality, co-incided sharply with the delivery path of the spray planes, leaving the mangrove areas adjacent to the spray path largely unaffected. All species subjected to the spray were affected, although those species which possess the capability of coppicing (e.g., *Avicennia*) often regenerated following a single light spraying. In contrast, *Rhizophora apiculata* most frequently exhibited mortality following a single spray operation. Coupled with the fact that this species is the dominant one in Viet Nam (both in ecological and economic terms), the extraordinary impact of the herbicides arises mainly from the loss of this single species.

Research on the effects of herbicides on mangroves largely confirms the relatively high sensitivity of mangroves to herbicidal defoliants. For example, Truman (1961–1962) showed that *Avicennia marina* was extremely sensitive to relatively small concentrations of 2,4-D. The sensitivity of other mangrove species was later confirmed in studies which showed that *Rhizophora mangle* could take up and translocate 2,4-D and picloram to various plant parts and that it induced the breakdown of cell walls in both roots and leaves (Walsh *et al.*, 1973; 1974). That the death of mangroves is due to the loss of meristematic tissues was confirmed by Lugo & Snedaker (1974), who showed that the complete mechanical (i.e., hand-picked) removal of both leaves and buds does not prevent reserve meristematic tissues from regenerating foliage. Thus, in the Vietnamese situation, the failure of many of the species, particularly among the Rhizophoraceae, to recover suggests that the reserve meristematic tissues were killed as well.

In conclusion, although many scientific questions remain, it is clear that the wartime herbicide operations had a significant and apparently lasting impact on the mangroves of Viet Nam.

III. Marine ecosystems

The marine ecosystems of interest in South Viet Nam are associated with the major mangrove areas. These complex systems are situated on deltaic landforms (e.g., the Mekong River delta and Saigon River delta [Rung Sat] areas) derived largely from alluvial silts and marine sands. Although they are subject to the actions of marine waves and currents, they also receive fresh water from the inland watersheds (catchments), a phenomenon that gives the water component an estuarine character (i.e., moderate to low salinity). As a result of the continual hydrologic activity, these areas tend to be highly intersected with tidal channels of various sizes which are linked to streams originating in the upland areas. They change over time, along with corresponding changes in the mangrove vegetation and marine organisms (Snedaker, 1982; Thom, 1982). Their interlinking water-ways and open areas are noted for a high productivity of estuarine and marine plants and animals which in turn is related to the high productivity of the man-groves, partially expressed in the production and export of leaf litter or detritus (Snedaker & Brown, 1982). For example, there exists a strong correlation between tropical estuarine environments dominated by mangrove vegetation and the commercial yields of penaeid shrimp (Martosubroto & Naamin, 1977; Turner, 1977).

In addition to the variety of roles of the mangroves and their detritus in the perpetuation of fishery yields (Snedaker, 1978), these coastal marine ecosystems are also characterized by fisheries that are supported by planktonic and benthic algal food webs. Although fishery data that pre-date the Second Indochina War are incomplete and somewhat vague, they indicate that South Viet Nam's fisheries were certainly being domestically and commercially exploited, and that they made a direct contribution to the country's economy (Brouillard, 1970; Shindo, 1973). It appears that Viet Nam has some 245 000 marine fisher-men and that all of the catch is locally consumed, mostly as fresh fish, but also as dried fish or as fish sauce (Westing, 1982, page 375). There are, however, no useful comparative data available that indicate the fishery yields obtained for a given level of fishing effort. Available regional data for the Gulf of Thailand and the South China Sea are not really suitable for extrapolation to the Viet Nam situation.

During the wartime spray operations over the mangrove areas, herbicides unavoidably entered into the waters that form the tidal channels of the affected areas. In addition, surface-water run-off carrying sediments and the relatively insoluble herbicides presumably continued to contaminate these bodies of water for considerable periods following the spray runs. There is, therefore, a good probability that some of the water area receiving direct and indirect loadings of herbicide exceeded the equivalent loadings of sprayed forest areas because of water run-off, erosion, tidal mixing, and the bed-load transport of contaminated sediments.

In comparison with the evidence for the forested environments, there is a paucity of direct evidence on the probable effect of herbicides on the estuarine and marine biota. Sylva & Michel (1974) conducted a sampling study of the Rung Sat estuarine areas, but were essentially unable to relate the observed effects directly to chemical toxicity of the herbicides. Their conclusions indicate that the major effects were probably attributable to the mortality of the man-groves and the concomitant increase in sediment erosion and water turbidity as well as to the sudden introduction of organic detritus from decomposing mangroves, followed by a lack of detritus in the years thereafter. Sylva & Michel did, however, acknowledge that herbicides introduced into these waters probably entered the food webs of the estuarine biota. Although Sylva & Michel evaluated the available information on fishery yields, they were unable to state whether or not the reported changes were due to herbicides and their indirect effects or to the overall effects of the war on the fishing 'effort' required to maintain a sustained annual harvest. They noted, for example, that the biological productivity appeared "to be sufficiently high" (but below the optimum) and that the spawning activity of fish was reduced, but not stopped.

In general, the equivocal observations and interpretations of pollutant effects in aquatic and marine ecosystems are typical of a generic problem. In the absence of direct evidence of an abundant mortality (dead plants and animals) or experimental data on the effects of herbicides on the Vietnamese marine biota, interpretations have to be based on indirect evidence and inference. Among the commonly cited inferences is the decrease in local fishery yields, although the available data base is too weak to support conclusions on a cause/effect relationship (see also Westing, 1982, page 375).

IV. Aquatic ecosystems

The aquatic ecosystems of Viet Nam consist of streams and lakes of various sizes and characteristics which are scattered throughout areas of different vegetational types. Like the marine ecosystems, the aquatic habitats were not specifically selected as targets for defoliant spray operations, but nevertheless received both direct spray and spray drift from operations over contiguous terrestrial vegetational types. In addition, they are presumed to have received herbicide residues from watershed run-off, especially during seasonal periods of heavy rainfall. Because of the relatively low water solubility of the herbicides, their indirect input into the aquatic ecosystems via run-off is associated primarily with particulate materials such as the products of erosion. Because increased precipitation run-off and surface soil erosion are common results of watershed deforestation, it can furthermore be presumed that the aquatic ecosystems eventually received a disproportionate loading of herbicides.

Based on the knowledge of fate and effects of herbicides in aquatic ecosystems, several probable consequences could be expected:

1. The submergent and emergent aquatic plants would experience decreased growth and increased mortality.

2. The herbicides would be concentrated in the benthic sediments and on the surfaces of plants and animals.

3. The more persistent herbicides could in time be transported downstream as part of the sediment bed load entering the coastal regions.

4. The herbicides in non-flowing habitats would tend to become concentrated in the sediments where eventual breakdown would more or less slowly reduce their abundance.

The 2,4-D would be broken down and detoxified in a relatively short period of time, approximating 120 d (Faust, 1964), whereas the more persistent compounds, such as dioxin, are variously reported to persist for extended periods measured in years (Westing, 1982, page 368). Edwards (1970), however, suggests that in general herbicides (as opposed to insecticides) do not persist for long periods in water and that their effects tend to be "immediate" rather than long-term.

Thus, there is essentially no valid information available on Viet Nam that would permit specific conclusions on the long-term consequences to aquatic ecosystems of the wartime use of herbicides. The absence of data, however, does not exclude the possibility that there are unrecognized long-term effects (see, e.g., Yên & Quýnh, paper 5.C).

V. Interpretation of the present situation

Erosion and sedimentation

Owing to the wartime loss of vegetational cover—that is, to the loss of the sediment stabilizing influence of the mangroves and the aquatic littoral vegetation—surface sediments and soils remain in a continuing state of erosion. The evidential base is extensive and includes comments and observations made during and after the war. The most notable piece of evidence is the abnormally high turbidity of the estuarine waters within and surrounding the mangrove areas. It is not known, however, to what extent the observed increase in turbidity is the result of local erosion within the mangrove ecosystem or of the delivery of eroded materials from the upstream watersheds in those forested areas that were sprayed. It is a well-known fact that the abuse of watersheds, primarily through the destruction of forest cover, results in upstream erosion and downstream sedimentation, accompanied by elevated levels of turbidity in the receiving waters. Irrespective of the source of the sediments, turbid water, high in silt and clay, has a deleterious effect on primary (green-plant) productivity and on the estuarine and near-shore filter feeders. Furthermore, the deposition and shoaling of sediments have an effect on local drainage and circulation patterns that can lead to a change in local salinity gradients (and also to limitation of access by the larger sizes of coastal vessels).

In the upland forested areas of the watersheds, vegetational communities are recovering (although not necessarily in a desired form) which has the effect of stabilizing the soil sediment and reducing the rate of erosion. It is expected that this watershed recovery process will continue and that the corresponding amounts of sediment that are delivered into the coastal zone will diminish proportionately. Within the mangrove forest areas, the relatively rapid recovery of the stream and tidal channel banks will minimize *in situ* erosion. In the mangrove areas where erosion has taken place, it is expected that the erosion phenomenon will accelerate the re-vegetation process owing to the exposure of reduced sediments and to the lowering of land elevation relative to surface water elevation. This process will assist in the removal of any persistent herbicides in the surface sediments and will promote the beneficial influence of surface water circulation and flushing. In the newly shoaled areas, developed from the deposition of the products of erosion, it is expected that the shoals will eventually be colonized by a variety of benthic plants and animals, constituting a new ecological system. Whereas the natural recovery of the upland and lowland ecosystems will reduce the rate of erosion, the deposited sediments may be presumed to retain some residual fraction of the more persistent water-insoluble herbicides. However, there is no evidence to confirm the presence of toxic materials in the areas of accretion, nor is there confirmatory evidence that any of the various colonizing forms are inhibited or affected by toxic materials.

Based on the observed recovery of the sediment-stabilizing plant communities, both in upland and lowland areas, it appears that no practical corrective action is warranted. However, poor inland land-use practices (e.g., shifting slash-and-burn agriculture, exploitative forest cutting) would, unless prevented, continue to contribute to soil erosion and deposition in the downstream and coastal habitats. It is also recommended that samples be taken from areas of new sediment accretion to determine whether or not they contain persistent toxic materials that may influence the recovery of the estuarine and marine biota. This should be considered a priority action since some benthic plants and animals constitute part of the food web that leads to humans.

Herbicide persistence

There is concern within Viet Nam that many continuing ecological problems are the direct result of herbicides and their residues persisting in the environment (specifically of Agent Orange and its dioxin contaminant). Among the commonly cited ecological problems are: the reduced numbers or absence of individuals of formerly common species; morphological anomalies among certain aquatic species; and the failure of plant communities to regenerate and recover their former status.

The very limited laboratory analyses of residual herbicides and their breakdown products in aquatic or marine sediments indicate that most do not persist at concentrations that warrant concern (Edwards, 1970). Likewise, the meagre data do not suggest that these materials are biologically accumulated or magnified to a significant degree. There are no corresponding data for the Viet Nam situation, thus making it difficult to arrive at a viable conclusion. The absence of data, however, should not be interpreted to mean that no problem exists, but only that insufficient studies have been undertaken to identify any existing problem or to assign a responsible cause to an observed effect. For example, analysis of recently collected sediment from the Rung Sat mangrove area reveals the continued presence of trace levels of dioxin more than a decade after its application (Olie, paper 9.B). However, the sample is not well documented and thus the information is not readily suitable for extrapolation or for estimating long-term consequences.

Until documented samples of plants, animals and sediments are taken and analysed based on a rigorous scientific protocol, it will not be possible to determine the extent and magnitude of the concentration of persisting herbicides, their contaminants, and breakdown products. Moreover, although results may show correlations between the persistence of herbicides and specific ecological problems, it must be recognized that even strong correlations are not by themselves proof of a cause/effect relationship.

Mangrove propagules

The extensive spraying of the mangrove forest resulted in an almost total loss of mature seed/propagule-bearing trees in the sprayed areas. The major consequence has been that the regeneration of the forest to its former species assemblage and dominance hierarchy has been largely prevented. With the exception of manual planting of the preferred species (*Rhizophora apiculata*), regeneration has been dominated by 'weed' species such as atypical grasses, the fern *Acrostichum aureum*, the palm *Phoenix paludosa*, several species of vines, and lesser valued mangroves such as *Ceriops*. Because of the inadequate natural regeneration of the forest, the former economic value has not been re-attained, nor is it likely to be for several decades. Indeed, estimates made by the US National Academy of Sciences indicated that forest re-establishment and recovery, presumably to its former state, would take approximately 100 years (Lang *et al.*, 1974).

Natural regeneration is, in fact, now taking place along stream and tidal channel banks that are well flushed during normal tidal cycles. Although the preferred species are not regenerating, the new vegetation does have an important initial ecological role in re-establishing the former habitat. In areas that are re-vegetating, even by grasses, the presence of the pioneer vegetation serves to trap and hold seeds and propagules, permitting their establishment and development (Ross, 1974). The colonizing vegetation induces changes in soil properties and plant community composition that favour successional development. Given sufficient time, it appears that natural successional processes will result in a viable mangrove forest community over much of the deforested habitat. Based on observations made in January 1983, this author estimates that a closed canopy forest will be established within another 10–15 years in those areas that are adequately flushed or otherwise suitable for mangrove colonization. This estimate is close to the 20-year recovery period that had been estimated by Tschirley (1969) during the war. The naturally rapid colonization of mangroves and mangrove associates along stream and tidal channel banks is expected to benefit local fisheries through the continual production of detritus.

Although the deforested mangrove areas are undergoing a relatively rapid regeneration process, the dominant pioneer species are not those that the Vietnamese recognize as having domestic, utilitarian or commercial value. Two courses of action are thus indicated. First, efforts should continue to be made to replant areas with those species that are recognized as economically valuable. Indeed, post-war replanting protocols have already been implemented, and by 1981 some 22 000 ha were re-established in the form of plantations of *Rhizophora apiculata*. The planting protocols are similar to those employed in many other South-east Asian and Pacific countries where manual replanting frequently follows the harvesting of mangrove timber (Christensen, 1982). As a second course of action, the currently colonizing pioneer species should be evaluated for economic use. For example, *Excoecaria agallocha* is known from Bangladesh to have excellent pulping properties and might well be managed here as a source of pulpwood for paper and paper-product manufacturing. In either instance, management standards should be enforced to prevent exploitative harvesting of immature timber for firewood or charcoal; this appears to be a major problem in the Rung Sat area owing to its proximity to the Ho Chi Minh City metropolitan area.

Habitat domination by weed species

In certain deforested areas, the palm *Phoenix paludosa* (a natural associate of mangrove communities) has become dominant to the exclusion of the various mangrove species of importance. In the more inland areas where the interstitial soil salinity is lower (as in the rear mangrove areas), the fern *Acrostichum aureum* assumes a similar dominant invading role. The high density of individuals of these species at maturity prevents, or at least greatly retards, the natural successional take-over and eventual dominance by mangroves. Such areas thus become largely unsuited for exploitation.

Most of the areas being colonized by the weed species are characterized by degraded soils not easily colonized by mangroves. The *Phoenix* palm, for example, is able to colonize and thrive on soils that are compacted, saline, and experience extended periods of drying. Areas dominated by palm or fern weed species are difficult to reclaim unless the vegetation is destroyed by extensive chopping and/or burning during periods of dry weather. In the absence of such control procedures, these areas can be expected to remain dominated for some time by species with little or no economic or utilitarian value.

An assessment of the extent of the invasion by palm and fern should be undertaken to determine whether or not corrective or control action is warranted. Palm and fern weed invasion problems are common throughout the region; and only as part of a broader silvicultural management plan should efforts be undertaken to eradicate them. The *Phoenix* palm usually becomes established in South-east Asia on drier soils that are not a preferred habitat for *Rhizophora* forests. If the total area of this weed species is small, or if its rate of invasion of new areas is minimal, it may not be practical to reclaim and replant these areas until all other immediately suitable areas have been re-established, either naturally or through planting.

Barren and degraded mangrove soils

As a result of the extensive deforestation of mangrove areas, the exposed soils have deteriorated in fertility, and large areas are now barren (see, e.g., Huây & Cu, paper 4.B). The altered soil characteristics make the land unsuitable for establishing new forest plantations.

Prior to the Second Indochina War, some 2 or 3 per cent of the Rung Sat mangrove forest area was classified as barren in the sense that the areas had no vegetation. By the end of the war, the amount had increased to 35 per cent (Ross, 1974). Various estimates have been made of the soil fertility and physical characteristics of the barren areas, and these uniformly show that exposure has led to a variety of significant changes in structure and fertility (Zinke, 1974; also paper 4.C), although the fertility levels appear to have remained generally suitable for mangrove growth and development. In general, it appears that the barren areas are those that originally were marginal in terms of vegetational growth and development. The majority of these areas are at the fringes of the mangrove zone and in the inland areas of the deltaic islands forming the Rung Sat area. Typically, these kinds of marginal habitat are indicative of the general evolution of mangrove-dominated land-forms, which results in the interiors of islands and the higher elevational fringes of mangrove zones becoming barren (Cintron *et al.*, 1978; Thom, 1982). It thus appears that the herbicidal destruction of these marginal areas magnified and accelerated the natural geomorphic process. As a result, the majority of the barren area is probably unlikely to support the regeneration of either mangrove forest or perhaps even a viable vegetational complex of other species.

Because of the total size of these barren areas and the poor prognosis for natural vegetational regeneration, it is recommended that they be evaluated for alternative uses such as rice cultivation or mariculture. Such evaluation must include an assessment of the potential for development of acid sulphate soil conditions which would limit any form of productive use. Owing to the current economic conditions in Viet Nam, a priority evaluation should be made of the potential of these areas as sites for brood and grow-out ponds for the culture of shrimp, a product that has a high value in domestic and international trade. In the more inland areas where there is more fresh water and lower soil salinities, the preferred crops might include freshwater prawns, carp, tilapia and milkfish.

Loss and/or deformation of aquatic species

In the inland aquatic habitats, there has been an absolute reduction in the number of species (Yên & Quýnh, paper 5.C). Among the freshwater algae, observations have been reported on the increased incidence of a variety of morphological aberrations (Nguyên Van Tuyên, University of Hanoi, unpublished). The findings of these two studies—made in the A Luói valley, Binh Tri Thien province, about 100 km west of Danang—are well documented and indicate that the aquatic habitats have undergone a significant change. The major cause of the decline in species and the high incidence of morphological aberrations in algal growth forms are attributed to the heavy use of herbicides in the watersheds surrounding the aquatic systems.

Although there is a good correlation between the herbicide spraying of inland watersheds and a variety of ecological problems in the contiguous aquatic systems, there is no evidence that the problems are wholly attributable to the persisting toxicity of herbicides or their residues. The A Luói valley (124 000 ha, 86 per cent forested) was subjected to heavy bombing and herbicide spraying to such a degree that more than a decade later the area was reported to remain devastated. In the context of such massive abuse of a watershed, it is not unreasonable to expect an equally massive change in the local fauna and flora, among them the components of the included aquatic ecosystems. However, the variety and magnitude of the abusive actions preclude assigning the effects simply to the herbicides or their residues.

Two courses of actions are warranted. First, statistically controlled samples should be taken and analysed to determine whether or not herbicides or their residues persist in the aquatic habitats within the watershed or in its drainage and run-off waters. Helpful in this regard might be the use of insect indicator species, which could help in identifying areas likely to contain persisting toxic materials. This should be followed by controlled experiments to induce the observed morphological aberrations in the same species of algae. Next, evaluations should be made to determine the most reasonable actions required in order to re-establish a closed canopy forest dominated by the former (or suitable new) species. Efforts to accelerate the natural recovery will, over time, provide a habitat for recolonization by the former aquatic fauna and flora, assuming, of course, that a reservoir of appropriate species is present in the general region.

Reduction in marine fishery yields

Immediately following the spraying and subsequent mortality of the coastal mangrove forests, fishery yields are reported to have been slightly higher (owing perhaps to a pulse input of detritus), but then to have fallen significantly lower than what is to be expected for a productive coastal and marine environment (Phan Nguyên Hong, Teachers Training College, Hanoi, unpublished). Estuarine habitats are noted for their highly productive fisheries, maintained in part by reduced salinity regimes, diversity of physical habitats, and high rates of primary productivity (including the production of mangrove detritus). In addition, it is not uncommon for many marine species, both demersal (bottom dwelling) and pelagic (open ocean), to spend part of their life cycle in estuarine environments because of the nurturing characteristics of the mangrove/estuarine habitat. The loss of so much mangrove-dominated forest area and the changes in the physical environment, including the possibility of persistent herbicide contamination, are considered to be responsible for the decline in fishery yields. The decline in yields cannot be attributed to exploitative over-fishing, because following the war, many Vietnamese emigrated to other countries using the most convenient means of transport: local fishing vessels. Although anecdotal, it was suggested that the total number of fishing vessels declined by 60–70 per cent following the war (see also Westing, 1982, page 375).

The currently available data base on fishery yields and fishing practices unfortunately is inadequate as a basis for drawing any conclusion regarding the status of fish populations. In addition to the probable effect of habitat destruction, lowered yields—if real—also appear to be correlated with restricted fishing activity, including inadequate or poorly maintained vessels and fishing gear. Furthermore, intensive fishing in local coastal waters, as opposed to offshore marine areas, could result in overfishing (and reduced yields) in the former and underfishing (and reduced yields) in the latter. The reported reductions in yield therefore seem to be attributable as much to infrastructural problems as to environmental problems.

Because of the importance of fisheries in the domestic economy of Viet Nam, it is urgent that an assessment be made of the fishing industry *sensu lato*. In addition to attempting to make evaluations based on absolute yield, fishery data should be obtained in a form that allows the expression of yield as a function of gear-normalized fishing effort in both inshore and offshore waters. Only after such information is evaluated can truly corrective plans be developed for either improving the fishery infrastructure or working towards an enhancement of coastal habitat conditions.

References

Brouillard, K. D. 1970. *Fishery Development Survey: South Vietnam*. Saigon: US Agency for International Development, 41 pp.

Buckingham, W. A., Jr. 1982. *Operation Ranch Hand: The Air Force and Herbicides in Southeast Asia 1961–1971*. Washington: US Air Force Office of Air Force History, 253 pp.

Christensen, B. 1982. *Management and Utilization of Mangroves in Asia and the Pacific*. Rome: FAO Environment Paper No. 3, 160 pp.

Cintron, G., Lugo, A. E., Pool, D. J. and Morris, G. 1978. Mangroves of arid environments in Puerto Rico and adjacent islands. *Biotropica*, Washington, **10**: 110–121.

Cruz, A. A. de la. 1979. *Functions of Mangroves*. Bogor, Indonesia: Southeast Asian Ministers of Education Organization, Regional Center for Tropical Biology, BIOTROP Special Publication No. 10, 225 pp.: pp. 125–138.

Edwards, C. A. 1970. *CRC Persistent Pesticides in the Environment*. London: Butterworths, 78 pp.

Faust, S. D. 1964. Pollution of the water environment by organic pesticides. *Clinical Pharmacology & Therapeutics*, St Louis, **5**: 677–686.

Lang, A. *et al.* 1974. *Effects of Herbicides in South Vietnam. A. Summary and Conclusions*. Washington: National Academy of Sciences, [398] pp. + 8 maps.

Lugo, A. E. and Snedaker, S. C. 1974. Ecology of mangroves. *Annual Review of Ecology & Systematics*, Palo Alto, Cal., **5**: 39–64.

Martosubroto, P. and Naamin, N., 1977. Relationship between tidal forests (mangroves) and commercial shrimp production in Indonesia. *Marine Research in Indonesia*, Jakarta, **1977**(18): 81–86.

Maurand, P. 1943. *Indochine Forestière*. Hanoi: Imprimerie d'Extrême-Orient, 252 pp.

Ross, P. 1974. *Effects of Herbicides in South Vietnam. B[14]. Effects of Herbicides on the Mangrove of South Vietnam*. Washington: National Academy of Sciences, 33 pp.

Shindo, S. 1973. *General Review of the Trawl Fishery and the Demersal Fish Stocks of the South China Sea*. Rome: FAO Fisheries Technical Paper No. 120, 49 pp.

Snedaker, S. C. 1978. Mangroves: their value and perpetuation. *Nature and Resources*, Paris, **14**(3): 6–13.

Snedaker, S. C. 1982. Mangrove species zonation: why? In: Sen, D. N. & Rajpurohit, K. S. (eds). *Tasks for Vegetation Science. II. Contributions to the Ecology of Halophytes*. Hague: Dr W. Junk, 272 pp.: pp. 111–125.

Snedaker, S. C. and Brown, M. S. 1982. Primary productivity of mangroves. In: Mitsui, A. & Black, C. C. (eds). *CRC Handbook of Biosolar Resources. I(2). Basic Principles*. Boca Raton, Fla.: CRC Press, 657 pp.: pp. 477–485.

Sylva, D. P. de and Michel, H. B. 1974. *Effects of Herbicides in South Vietnam. B[15]. Effects of Mangrove Defoliation on the Estuarine Ecology and Fisheries of South Vietnam*. Washington: National Academy of Sciences, 126 pp.

Thom, B. G. 1982. Mangrove ecology: a geomorphological perspective. In: Clough, B. F. (ed.). *Mangrove Ecosystems in Australia: Structure, Function and Management*. Canberra: Australian Institute of Marine Science, 302 pp.: pp. 3–17.

Truman, R. 1961–1962. Eradication of mangroves. *Australian Journal of Science*, Sydney, **24**: 198–199.

Tschirley, F. H. 1969. Defoliation in Vietnam. *Science*, Washington, **163**: 779–786.

Tung, Thái Công. 1967. *Natural Environment and Land Use in South Viet-Nam.* 2nd ed. Saigon: Republic of Viet Nam Ministry of Agriculture, 156 pp. + 3 maps.

Turner, R. E. 1977. Intertidal vegetation and commercial yields of penaeid shrimp. *Transactions of the American Fisheries Society*, Washington, **106**: 411–416.

Walsh, G. E., Barrett, R., Cook, G. H. and Hollister, T. A. 1973. Effects of herbicides on seedlings of the red mangrove, *Rhizophora mangle* L. *BioScience*, Washington, **23**: 361–364.

Walsh, G. E., Hollister, T. A. and Forester, J. 1974. Translocation of four organochlorine compounds by red mangrove (*Rhizophora mangle* L.) seedlings. *Bulletin of Environmental Contamination and Toxicology*, New York, **12**(2): 129–135.

Westing, A. H. 1976. In: SIPRI, *Ecological Consequences of the Second Indochina War.* Stockholm: Almqvist & Wiksell, 119 pp. + 8 pl.

Westing, A. H. 1982. Environmental aftermath of warfare in Viet Nam. In: *World Armaments and Disarmament, SIPRI Yearbook 1982.* London: Taylor & Francis, 518 pp.: pp. 363–389.

Zinke, P. J. 1974. *Effects of Herbicides in South Vietnam. B[19]. Effect of Herbicides on Soils of South Vietnam.* Washington: National Academy of Sciences, 39 pp.

Chapter Six
Cancer and Clinical Epidemiology

6.A. Cancer and clinical epidemiology : Symposium summary[1]

Luong Tan Truong *et al.*[2]

Institute of Cancer, Ho Chi Minh City

This summary report examines epidemiologically the long-term effects on human cancer and other clinical sequelae of the herbicides applied in South Viet Nam during the Second Indochina War of 1961–1975. Of particular concern here is the dioxin contaminant of Agent Orange, the major herbicide that was employed.

Dioxin is one of the most toxic organic compounds known, producing a wide range of organ dysfunctions, metabolic dysfunctions, foetotoxicity, teratogenicity and carcinogenicity at the ng/kg to μg/kg (ppt to ppb) range. There is a general consistency between the pattern of chronic toxicity induced in animals by dioxin and dioxin-contaminated chlorophenolic compounds and those observed in exposed human populations. Such toxicity includes: (*a*) chronic hepatitis (liver inflammation); (*b*) disturbances in immune function; (*c*) disturbances in lipid and porphyrin pigment metabolism; and (*d*) neurological abnormalities, sometimes associated with toxic neurasthenia (ill-defined weakness). Studies by Tung (1973) on Vietnamese populations exposed to herbicides during the Second Indochina War have produced suggestive evidence of an excess of primary liver cancers and other evidence of chronic toxicity. A series of Swedish epidemiological studies, confirmed by more recent US mortality studies, have demonstrated an excess of soft-tissue sarcomas[3] in groups occupationally exposed to chlorophenoxy herbicides and chlorophenolic

[1] Summary report of a Working Group of the International Symposium on Herbicides and Defoliants in War: the Long-term Effects on Man and Nature, Ho Chi Minh City, 13–20 January 1983 (appendix 3).
[2] The Working Group consisted of N. S. Antonov, Doàn Thúy Ba, Tôn That Bach, H. Carpentier, Nguyên Dinh Dich, Z. Dienstbier, J. H Dwyer, S. S. Epstein (Rapporteur), K.-R. Fabig, R. R. Gavalda, C. R. Jerusalem, Pham Duy Linh, My Samedy, Pham Hoàng Phiêt, N. S. Scharager, Pham Bieu Tam, C. Thammavong, C. J. M. van Tiggelen, Dô Thuc Trinh, Le The Trung, Luong Tan Truong (Chairman), Nguyên Anh Tuong, Dô Dúc Vân (Vice-chairman) and Nguyên Van Van.
[3] 'Soft-tissue sarcoma' is a fleshy malignant tumour of the soft somatic tissues (Morton & Eilber, 1982). The soft somatic tissues, which make up more than 50 per cent of the body weight, include the fibrous and adipose connective tissues, blood vessels, nerves, smooth and striated muscles, fascia (ligaments, tendons and so on) and their synovial structures (sheaths and so on), and lymphatic structures. However, cancers of the lymphatic system—Hodgkin's disease (Rosenberg, 1982), Burkitt's lymphoma (Ziegler, 1982) and other malignant lymphomas (DeVita *et al.*, 1982)—are not usually subsumed under the term 'soft-tissue sarcoma'.

compounds. Chloracne is *not* an obligate effect of dioxin exposure in either sensitive animal species or humans.

Vietnamese studies

Morbidity studies of civilians in Tay Ninh province (in the former Military Region III, War Zone C) and in Ben Tre province (in the former Kien Hoa province, Military Region IV) (Trinh, paper 6.B) as well as of Vietnamese veterans in the North have demonstrated consistent and strong associations between wartime herbicide exposure and chronic neurasthenic symptoms. Two preliminary case-control (case-referent) studies of primary liver cancer were reported. A case-control study of primary liver cancer in Hanoi demonstrated a strong association with herbicide exposure (Vân, paper 6.C). Another similar study at Cho Ray Hospital in Ho Chi Minh City with a limited number of cases gave evidence of a slight excess of risk of liver cancer in exposed persons, but this association was not large enough to achieve statistical significance.

These Vietnamese studies have established suggestive evidence of an association between wartime herbicide exposure and chronic toxic effects, including neurasthenic symptoms and primary liver cancer. It is planned to expand these studies with particular reference to the following four points: (*a*) definition of past and present exposure to toxic herbicides, including dioxin levels, from direct and indirect sources; (*b*) methodological considerations, including the need for larger sample sizes, random sampling, the use of multiple controls, and avoidance of reporting bias; (*c*) incorporation of objective clinical and laboratory studies, such as associations between chronic neurasthenic symptoms and disturbances in nerve conduction velocity and in lipid metabolism and porphyrin metabolism; and (*d*) study of the role of hepatitis-B in studies of the association between primary liver cancer and exposure to toxic herbicides.

The Working Group recognizes the major problems in conducting complex epidemiological studies of this type, even under ideal conditions, and congratulates its Vietnamese colleagues for their scientific contributions under difficult conditions.

Recommendations

Although primary consideration here has been directed towards Viet Nam, the Working Group recognizes the existence of similar problems and needs in Laos and Kampuchea as a result of the Second Indochina War. Greatly expanded initiatives should be developed in the following five general areas: (*a*) collaborative programmes that are based jointly in Vietnamese and foreign laboratories; (*b*) visiting consultant programmes that involve foreign scientists with work in Viet Nam; (*c*) scholarship programmes that allow young Vietnamese scientists to receive specialized training in foreign countries; (*d*) the development of standardized protocols for epidemiological, clinical and laboratory studies; and (*e*) foreign reference centres for specialized purposes such as dioxin analysis and histopathology review.

Attempts should be made to integrate such initiatives with world-wide studies on groups occupationally exposed to dioxin and dioxin-contaminated chlorophenoxy compounds, including foreign veterans of the Second Indochina War. Such initiatives should be developed in parallel with programmes to improve the overall public health and nutritional status of the Vietnamese population. Specific recommendations for collaborative programmes include the following five areas: (a) expanded case-control studies designed to investigate the relationship in standardized populations between past exposure to the toxic herbicides and present disease; (b) the same studies designed to also investigate associations between subjective disease and objective clinical and laboratory findings; (c) the same studies additionally designed to investigate the relation between such associations and present levels of dioxin in soil, water, and vegetation in areas that had been subjected to wartime spraying; (d) retrospective case-control studies on soft-tissue sarcomas; and (e) subject to available resources, long-term prospective epidemiological studies on exposed Vietnamese populations.

The Working Group recognizes that all recommendations are meaningless in the absence of a workable plan for implementation. The following five recommendations for practical action are therefore proposed:

1. Funding should be sought to support further research, diagnosis, and treatment of the effects of the wartime exposure to herbicides in Viet Nam, Laos and Kampuchea.

2. Practical mechanisms for scientific collaboration should be established immediately. In particular, these mechanisms should include international scientific commissions or committees for collaborative research.

3. Every effort should be made by the participants in this Symposium to increase the availability of medical supplies to Vietnamese, Laotian and Kampuchean researchers. Similar efforts should be made with respect to scientific journals, laboratory reagents and laboratory equipment.

4. Research concerning the treatment of exposed persons should be part of the overall research programme.

5. The World Health Organization (WHO) should be approached concerning the expansion of the dioxin project of its International Agency for Research on Cancer (IARC) in order to incorporate and support research on the effects of herbicides in Indochina.

References

DeVita, V. T., Jr, Fisher, R. I., Johnson, R. E. and Berard, C. W. 1982. Non-Hodgkin's lymphomas. In: Holland, J. F. & Frei, E., III (eds). *Cancer Medicine*. Philadelphia: Lea & Febiger, 2465 pp.: pp. 1502–1537.
Morton, D. L. and Eilber, F. R. 1982. Soft tissue sarcomas. In: Holland, J. F. & Frei, E., III (eds). *Cancer Medicine*. Philadelphia: Lea & Febiger, 2465 pp.: pp. 2141–2157.
Rosenberg, S. A. 1982. Hodgkin's disease. In: Holland, J. F. & Frei, E., III (eds). *Cancer Medicine*. Philadelphia: Lea & Febiger, 2465 pp.: pp. 1478–1502.
Tung, Ton That. 1973. [Primary cancer of the liver in Viet Nam.] (In French) *Chirurgie*, Paris, **99**: 427–436.
Ziegler, J. L. 1982. Burkitt's lymphoma. In: Holland, J. F. & Frei, E., III (eds). *Cancer Medicine*. Philadelphia: Lea & Febiger, 2465 pp.: pp. 1537–1546.

6.B. Long-lasting morbid effects following herbicidal attack

Dô Thuc Trinh[1]

College of Medicine, Hanoi

I. Introduction

Large areas of forest and agricultural lands in South Viet Nam were attacked with herbicides during the Second Indochina War, especially from 1966 to 1969. As a result many people were directly or indirectly exposed to these agents, of which Agent Orange (with it dioxin contaminant) made up more than 60 per cent in terms of volume sprayed.

In the present study the question was addressed as to whether wartime exposure to the herbicides sprayed has had a persistent effect on subsequent human morbidity patterns.

II. Materials and methods

The study area was four communes of similar geography, sociology, and economy within Giong Trom district, Ben Tre province (in the former Kien Hoa province), approximately 75 km south-south-west of Ho Chi Minh City. A total of 558 subjects was selected, aged 16 and older, of whom 227 were male and 331 female (table 6.B.1). Of these, 358 resided in the sprayed communes of Luong Phu, Luong Hoa, or Thuan Dien and had each been directly exposed (i.e., sprayed upon) one or more times between 1966 and 1968. The remaining 200 resided in the unsprayed commune of My Thanh and had never been directly exposed.

The study area was visited in 1982 by a group of medical doctors representing several specialties. Each of the 558 subjects was carefully interviewed and clinically examined in detail. Moreover, blood samples were collected for routine haematological testing from 324 of the subjects (264 exposed, 60 unexposed) and for liver-associated biochemistry from 138 subjects (73 exposed, 65 unexposed).

---

[1] Adapted by the editor from the author's Symposium presentation.

Table 6.B.1. Sex and age distributions of the subjects, 1982

Age class (years)	Age span (years)	Without past spray exposure[a]		With past spray exposure[b]		Combined age distribution (per cent)
		Male	Female	Male	Female	
16–25	9	25	32	22	21	18
26–45	19	35	54	62	87	43
46–60	14	13	29	57	66	30
61 and over		3	9	10	33	10
Total		76	124	151	207	100

[a] From My Thanh commune, Giong Trom district, Ben Tre province.
[b] From Luong Phu, Luong Hoa, and Thuan Dien communes, Giong Trom district, Ben Tre province.

III. Results

The most dramatic clinical finding was that chronic hepatitis (inflammation of the liver) was more than 10 times as prevalent among those subjects who had been directly exposed to the military herbicides (more than a decade previously) than among those who had not (table 6.B.2). Moreover, the interviews revealed that only 12 per cent of the exposed group found to suffer from chronic hepatitis did so prior to the spraying (i.e., at a pre-spray frequency of 8 per 1 000). The most seriously affected age class was 46–60. Clinical examination revealed no case of liver cancer among the 558 subjects. The blood serum of unexposed subjects contained the normal minimal amounts of the two enzymes glutamic-oxaloacetic transaminase and glutamic-pyruvic transaminase and of the protein(s) precipitated by thymol. Conversely, the serum of some of the previously exposed subjects contained elevated levels of these substances, a finding which is indicative of hepatic (liver) damage (in 11 per cent of the cases tested for the two enzymes and in 5 per cent of the cases tested for the protein moiety).

The frequency of chronic weakness of various sorts—both malnutrition asthenia and neurasthenia—was found to be about two or three times as great among the previously exposed subjects as among the unexposed subjects. This may be the result of the increase in various physiological diseases (table 6.B.2) or it could arise from the psychological stress of having been exposed to the chemicals (namely, through a fear of adverse effects to oneself and to one's offspring). A direct primary toxic effect of the sprayed chemicals may also have contributed to the effect.

A variety of routine blood tests was performed on both the previously exposed and unexposed groups, including red blood cell counts (both erythrocyte and reticulocyte counts), sedimentation speed, clotting time, lymphocyte (white blood cell) morphology, and haemoglobin content. No significant haematological difference could be discerned between the two groups. (The high frequency of clinical anaemia in both groups was confirmed by the blood tests.)

116

Table 6.B.2. Morbidity in Giong Trom district, Ben Tre province, 1982

Medical problem	Without past spray exposure[a] (No./1 000)	With past spray exposure[b] (No./1 000)	Change[c] (per cent)
Malnutrition asthenia (weakness)	50	128	+156*
Neurasthenia (ill-defined weakness)	35	126	+260*
Anaemia	50	84	+68
Angina pectoris (pain about the heart)	35	64	+83
Chronic rhinitis (nasal inflammation)	85	84	−1
Asthma	65	50	−23
Gastroduodenitis (stomach and small intestinal inflammation)	55	126[d]	+129*
Chronic hepatitis (liver inflammation)	5	67[e]	+1 240*
Periodontitis (mouth inflammation)	20	89	+345*
Menstrual disturbance	15	31	+107
Pruritis (skin irritation)	10	8	−20
Rheumatism	30	17	−43

[a] Based on 200 subjects (16 and older) in My Thanh commune.
[b] Based on 358 directly exposed subjects (16 and older) in Luong Phu, Luong Hoa, and Thuan Dien communes.
[c] Those changes indicated by an asterisk (*) were found by t test to be significant at the 1 per cent level or better.
[d] In 92 cases per thousand this problem appears from interviews to post-date the exposure to herbicide spraying.
[e] In 59 cases per thousand this problem appears from interviews to post-date the exposure to herbicide spraying.

IV. Conclusion

It seems clear that wartime exposure to the herbicides used can lead to a variety of long-lasting morbid effects among humans, including chronic weakness, gastro-intestinal problems, inflammation of the mouth, and, most prominently, liver damage. The possibility exists that the hepatic problems stem from the dioxin contaminant of Agent Orange. In this regard it must be noted that Vân (paper 6.C), on the basis of his cancer studies, has also suggested that the liver is the major target organ for dioxin.

6.C. Herbicides as a possible cause to liver cancer

Dô Dúc Vân[1]

Viet Duc Huu Nghi Hospital, Hanoi

I. Introduction

The Viet Duc Huu Nghi Hospital in Hanoi has in the past received about 1 000 new cancer patients each year. Dr Ton That Tung, former Chief of Surgery (and Director) of the Hospital, noted that the fraction of such patients who suffered from cancer of the liver (primary hepatic carcinoma) was approximately 3 per cent prior to the Second Indochina War, but that this fraction more than trebled during the war years. More precisely, the values reported by Tung (1973) for the six-year period 1955–1961 were 159 of 5 492 (for an annual rate of 29 per 1 000); and for the subsequent six-year period 1962–1968, 791 of 7 911 (for an annual rate of 100 per 1 000).

Tung suggested the possibility that the noted relative increase in liver cancer was the result of exposure to the military herbicides that were being employed in South Viet Nam, and, more specifically, to Agent Orange (which accounted for more than 80 per cent of these, in terms of active ingredients) and its dioxin impurity. The wartime spraying occurred essentially during the period 1962–1970, and was especially heavy from 1966 to 1969. Tung's suggestion found support in the findings of Buu-Hoï *et al.* (1972a; 1972b) that the liver was the main target organ of dioxin when administered to rats (see also Greenlee & Poland, 1979; Poland & Glover, 1974). Moreover, various reports in the literature indicate that dioxin exposure in rats can, indeed, lead to liver tumours and carcinomas (Kociba *et al.*, 1978; Van Miller *et al.*, 1977).

The present study investigated to what extent recent victims of liver cancer had a history of wartime exposure to the military herbicides.

II. Materials and methods

A total of 63 cancer or ulcer patients admitted to the Viet Duc Huu Ngai Hospital during the period January–September 1982 were selected for study. The sample was restricted to males 18 to 50 years old. Twenty-one of the subjects suffered from cancer of the liver (primary hepatic carcinoma) and the remaining

[1] Adapted by the editor from the author's Symposium presentation.

42 from either cancer of the stomach (gastric carcinoma, 8 cases) or ulcer of the subjacent small intestine (duodenal ulcer, 34 cases). Diagnosis was in each case established during surgery and verified by a pathological examination. (Five of the 21 hepatic cancer cases were found to also suffer from cirrhosis of the liver.)

Each subject was interviewed by two separate interviewers in order to determine whether, for how long, and where he had been in South Viet Nam during the war; and whether and to what extent he had been exposed during that time to the herbicides that had been sprayed. A subject was considered to have been exposed if he had been living, working or fighting in sprayed regions of South Viet Nam at the time of spraying or subsequently. Nine of the subjects were found to have been exposed in this sense for varying durations between 1966 and 1976, the minimum duration having been 8 months, the maximum 77, and the average 47. The remaining 54 subjects had not been exposed.

III. Results

Of the 21 subjects with liver cancer in 1982, 6 had been exposed to the wartime herbicides, that is, 29 per cent of the group. The average duration of such exposure had been 56 months (range, 24–77), occurring during the period between 1966 and 1976. Conversely, of the 42 subjects with stomach cancer or intestinal ulcer in 1982, only 3 had been exposed to the wartime herbicides, that is, 7 per cent of the group. The average duration of such exposure had been 27 months (range, 8–58), occurring during the period between 1969 and 1975.

Looked at in another way, of the 54 patients studied who had not been exposed to the wartime herbicides, 15 had liver cancer, that is, 28 per cent of the group. Conversely, of the 9 patients who had been exposed to the wartime herbicides, 6 had liver cancer, that is, fully 67 per cent of this latter group.

IV. Conclusion

The findings, although based on a very small sample, suggest that exposure by humans to the herbicides used during the Second Indochina War—and presumably their dioxin contaminant—results rather rapidly in a predisposition to develop cancer of the liver. It is of interest to note in this regard that Trinh (paper 6.B) has observed that wartime herbicide exposure has led to a high incidence of chronic hepatitis.

References

Buu-Hoï, N. P., Chanh, Pham-Huu, Sesqué, G., Azum-Gelade, M. C. and Saint-Ruf, G. 1972a. Enzymatic functions as targets of the toxicity of "dioxin" (2,3,7,8-tetrachlorodibenzo-p-dioxin). *Naturwissenschaften*, W. Berlin, **59**: 173–174.
Buu-Hoï, N. P., Chanh, Pham-Huu, Sesqué, G., Azum-Gelade, M.C. and Saint-Ruf, G. 1972b. Organs as targets of "dioxin" (2,3,7,8-tetrachlorodibenzo-p-dioxin) intoxication. *Naturwissenschaften*, W. Berlin, **59**: 174–175.

Greenlee, W. F. and Poland, A. 1979. Nuclear uptake of 2,3,7,8-tetrachlorodibenzo-*p*-dioxin in C57BL/6J and DBA/2J mice. *Journal of Biological Chemistry*, Baltimore, **254**: 9814–9821.

Kociba, R. J. *et al.* 1978. Results of a two-year chronic toxicity and oncogenicity study of 2,3,7,8-tetrachlorodibenzo-*p*-dioxin in rats. *Toxicology and Applied Pharmacology*, New York, **46**: 279–303.

Poland, A. and Glover, E. 1974. Comparison of 2,3,7,8-tetrachlorodibenzo-*p*-dioxin, a potent inducer of aryl hydrocarbon hydroxylase, with 3-methylcholanthrene. *Molecular Pharmacology*, New York, **10**: 349–359.

Tung, Ton That. 1973. [Primary cancer of the liver in Viet Nam.] (In French) *Chirurgie*, Paris, **99**: 427–436.

Van Miller, J. P., Lalich, J. J. and Allen, J. R. 1977. Increased incidence of neoplasms in rats exposed to low levels of 2,3,7,8-tetrachlorodibenzo-*p*-dioxin. *Chemosphere*, Elmsford, N.Y., **6**(9): 537–544 [also **6**(10): 625–632].

6.D. Cancer and clinical epidemiology : an overview

James H Dwyer and Samuel S. Epstein
State University of New York and University of Illinois Medical Center

I. Introduction

Large quantities of herbicides were dispensed from the air for hostile purposes in South Viet Nam during the Second Indochina War, which inevitably led to human exposure. Of the several herbicides involved, 2,4-D represented about 48 per cent of the total by weight and 2,4,5-T (with its dioxin contaminant) another 44 per cent. This survey has the purpose of reviewing the literature concerned with human malignancies and neurotoxic effects of exposure to these phenoxy herbicides; of summarizing a number of recent unpublished Vietnamese studies; and of considering needs for further research relevant to Viet Nam.

Since our direct knowledge on human health effects of exposure to phenoxy herbicides and their contaminants necessarily stems from non-experimental situations, there is usually greater uncertainty about such effects than those revealed by laboratory animal studies. Thus, causal inferences must be made with the clear understanding that some confounding variables may qualify the significance of the association between herbicide exposure and adverse health effects. Nevertheless, there is a substantial literature concerning the toxic clinical effects which have been associated with such exposure. It is additionally recognized that the dynamics of human exposure to herbicides during the Second Indochina War may have been sufficiently different from those of other large-scale exposures to have led to unanticipated clinical patterns.

II. Neurotoxicity of phenoxy herbicides and contaminants

2,4-D

Evidence of the human neurotoxicity of 2,4-D has accumulated since the late 1950s. Goldstein *et al.* (1959) reported on three cases (2 males, 1 female) of 2,4-D polyneuritis (inflammation of the spinal nerves) that developed a few hours after exposure during agricultural spraying. The symptoms were described as progressing through a period of days until pain, paraesthesia (abnormal sensations) and paralysis were severe. Disability was protracted, and recovery was incomplete even after a lapse of years. Further symptoms included asthenia (general weakness), nausea, numbness of the extremities, fasciculations (muscular twitching) and hyporeflexia (diminished functioning of the reflexes).

There have been more than a dozen subsequent case reports of neurotoxicity following acute exposure to 2,4-D, both cutaneous and oral, with the following findings (JRB Associates, 1981): asthenia (8 cases); hyporeflexia (8 cases); ataxia (muscular inco-ordination, 8 cases); hypaesthesia (diminished touch sensitivity, 6 cases); and myotonia (muscle rigidity or spasms, 7 cases). There exists confirmatory evidence of an analagous syndrome in laboratory animals (Danon *et al.*, 1978; Eberstein & Goodgold, 1979; Hill & Carlisle, 1947). Additionally, Elo & Ylitalo (1977; 1979) have found evidence in rats for involvement of the central nervous system. Thus, the convergence of the various case reports of human exposure to 2,4-D with the animal studies produces a convincing picture that this substance can be neurotoxic.

Subacute animal exposure studies with 2,4-D have produced uneven results. Drill & Hiratzka (1953) administered the 2,4-D orally to dogs almost daily for 13 weeks. Myotonia was observed, with severity ranging from mild ataxia to debilitating paralysis. However, Kay *et al.* (1965) administered various formulations of 2,4-D dermally to rabbits, which produced no change in either the gross or microscopic characteristics of the peripheral or central nervous system tissues. Similar negative findings have been reported for cattle (Palmer, 1963) and sheep (Palmer & Radeleff, 1964).

2,4,5-T

Exposures to 2,4,5-T are always confounded by concomitant exposure to its dioxin contaminant, the latter discussed independently below. There appears to be no report of neurotoxicity in animal studies using highly purified 2,4,5-T. The several reports of anorexia (loss of appetite), ataxia and muscle stiffness following oral administration have been carried out with unpurified (commercial) grades of 2,4,5-T (e.g., Drill & Hiratzka, 1953).

Dioxin

One episode of human exposure to dioxin is of particular interest (Oliver, 1975). It involved three male laboratory workers engaged in the synthesis of this compound. Eight weeks after exposure, one man developed chloracne (acne-like skin eruption due to contact with certain chlorinated compounds), with no apparent other evidence of toxicity, which subsided after 18 months. The second man also developed chloracne which cleared up within a year. But two years after exposure he developed neurotoxicity, experiencing such symptoms as fatigue, headaches, decreased concentration, irritability, abdominal pain, flatulence, hirsutism, blurred vision and ataxia, which subsided within six months. The third man developed no chloracne, but three years after exposure exhibited an inability to concentrate, diarrhoea, diminished sense of taste, visual problems, insomnia, thigh pains, oiliness of the skin and hirsutism; these symptoms also subsided in about six months.

Another instance of dioxin exposure resulted from the use of contaminated oil as a dust palliative in a horse arena (Beale *et al.*, 1977; Carter *et al.*, 1975; Epstein *et al.*, 1982; Kimbrough *et al.*, 1977). Numerous horses and other animals died; and several exposed humans, both children and adults, exhibited signs of acute dioxin intoxication. The human symptoms, which appeared within a week of exposure, included headaches, diarrhoea, haematuria (blood in the urine) and abdominal pain. A follow-up study five years later revealed no sign of neuro-intoxication.

One of the few studies dealing with the neurotoxic effects of dioxin in animals was carried out by Creso *et al.* (1978). They found that either oral or intraperitoneal (within the abdomen) administration of dioxin to rats led to signs of central nervous system dysfunction, including irritability, restlessness and increased aggression.

Mixed exposure

Between 1949 and 1976 there have been at least seven factory explosions during the production of trichlorophenol that resulted in the concomitant synthesis and release of dioxin (Hay, 1979; 1982; Huff *et al.*, 1980; JRB Associates, 1981). The two most commonly reported conditions following from these industrial accidents were chloracne and asthenia. Consistently reported as well were fatigue, irritability, and insomnia; also reported was loss of libido. These and the other symptoms which were noted are highly suggestive of neurotoxicity (see, e.g., Pocchiari *et al.*, 1979). It is important to note here that the chloracne was very probably overrated in these reports because many of the investigators focused only upon those workers who developed chloracne, in the mistaken belief that this was an obligate marker of dioxin exposure.

Zack & Suskind (1980) have carried out a follow-up study of 121 cases resulting from one of the above-noted factory accidents. On the basis of an analysis of the 32 subsequent deaths they reported that there was no evidence of excess mortality. However, re-analysis of these data by one of the present authors (SSE) suggests that they do not support this conclusion: it appears that there was, indeed, excess mortality from malignant lymphoma (cancer of the lymphatic system) and leukaemia.

There have also been 11 reported instances of industrial dioxin exposure involving improper procedures during the production of trichlorophenol that, however, did not culminate in an explosion (Huff *et al.*, 1980; JRB Associates, 1981). Here the symptoms described are similar to those noted above, although with the addition of gastro-intestinal disorders.

In a recent study, Singer *et al.* (1982) examined seemingly healthy workers in a factory that produced 2,4-D and 2,4,5-T. The nerve conduction velocity of 76 such workers was compared with that of 22 controls. Interestingly enough, 46 per cent of the workers exhibited a slowed velocity as compared with only 5 per cent of the controls. Moreover, there was some correlation found between length of employment and the magnitude of this reduction in nerve conduction velocity.

Vietnamese studies

Two recent Vietnamese studies can be cited that are of relevance to the neurotoxicity of phenoxy herbicides. Trinh (paper 6.B) investigated morbidity patterns in a region about 75 km south-south-west of Ho Chi Minh City. She compared 350 individuals who had been directly exposed during the Second Indochina War to herbicides (likely to have been phenoxy herbicides) with 200 who had not (although details of the basis of selection were not described). Trinh reported significant differences between the exposed and control groups for five medical problems: chronic hepatitis (liver inflammation), periodontitis (mouth inflammation), neurasthenia (ill-defined weakness), malnutrition asthenia and gastroduodenitis (stomach and small intestinal inflammation). The increase in chronic hepatitis was especially pronounced. The findings of Trinh are consistent with those reported above with the exception of the hepatitis, a condition which appears to be somehow peculiar to Viet Nam.

In the second study, Tôn Dúc Lang & Dô Dúc Vân (Viet Duc Huu Nghi Hospital, Hanoi, unpublished) used as their study group 432 male veterans of the Second Indochina War over 18 years old residing at the time of the study (1982) in rural areas in the vicinity of Hanoi (all such individuals who could be located in three communes, representing 85 per cent of the total). Each subject was given a physical examination by a group of medical doctors and interviewed by another such group. On the basis of the interviews each was categorized regarding wartime herbicide exposure as having experienced 'high' (117 subjects), 'low' (177 subjects), or 'no' (138 subjects) exposure. The following medical problems were reported to be more prevalent among the high-exposure group than among the no-exposure group (these differences all being statistically significant at the 5 per cent level): episodes of fatigue; loss of appetite; eyes sensitive to light; unexplained vomiting; erratic fits of rage; headaches; and periods of depression. Although the high-exposure group was older than the no-exposure group (72 per cent versus 39 per cent over 35), this can presumably be ignored since statistical analysis revealed no interaction between age and the listed medical problems.

III. Carcinogenicity of phenoxy herbicides and contaminants

The most convincing evidence of a causal link between phenoxy herbicide exposure and increased risk of malignancy in humans is that of Hardell and colleagues (Eriksson et al., 1981; Hardell, 1981; Hardell & Sandström, 1979; Hardell et al., 1981). These authors discovered that workers (agricultural and forestry sprayers) occupationally exposed to these compounds suffer from a markedly elevated incidence of soft-tissue sarcomas and malignant lymphomas[1].

---

[1] 'Soft-tissue sarcoma' is a fleshy malignant tumour of the soft somatic tissues (Morton & Eilber, 1982). The soft somatic tissues, which make up more than 50 per cent of the body weight, include the fibrous and adipose connective tissues, blood vessels, nerves, smooth and striated muscles, fascia (ligaments, tendons and so on) and their synovial structures (sheaths

The excess risk of soft-tissue sarcoma has been confirmed in so-called pooled-cohort occupational studies in the USA (Honchar & Halperin, 1981). These findings are also consistent with the results of a number of animal carcino-genicity studies (IARC, 1982, pages 238–243; Kociba *et al.*, 1978; Kouri *et al.*, 1978; Pitot *et al.*, 1980; Van Miller *et al.*, 1977).

Vietnamese studies

Another two recent Vietnamese studies can be cited, these of relevance to the carcinogenicity of phenoxy herbicides. Vân (paper 6.C) explored the possibility of wartime herbicide exposure in 21 recent cases of liver cancer (primary hepatic carcinoma). Of this group, 6 turned out to have been thus exposed during the Second Indochina War, a value considered to have been high enough (in comparison with a control group of 42) to suggest a cause–effect relationship. It should be mentioned that Vân's investigation had been prompted by the finding of Tung (1973) that the proportional incidence of primary hepatic cancers had risen during the war years. However, it must be noted that Vân's sample was small and his sampling procedure perhaps biased; and it is also not clear whether his procedures would have been able to discriminate between phenoxy herbicides and other possible confounding factors in the aetiology of the observed tumours (e.g., certain parasites, hepatitis-B virus, aflatoxin).

Pham Hoàng Phiêt (Cho Ray Hospital, Ho Chi Minh City, unpublished) recently carried out a similar study on 26 cases of primary hepatic cancer (and 52 controls), but in the view of the present authors with inconclusive results.

IV. Conclusion

There is sufficient evidence from outside Viet Nam to indicate that those persons heavily exposed to phenoxy herbicides and their contaminants are at increased risk regarding a number of long-lasting neuropsychological dys-functions and certain malignancies. The pilot studies from Viet Nam suggest the need for expanded public health and increased research attention to these matters in that country. It seems appropriate, moreover, that such efforts be supported by the international medical and scientific communities. There should be a particular focus in Viet Nam on investigations of primary hepatic carci-noma, soft-tissue sarcoma, malignant lymphoma and neuro-intoxication. The last should incorporate objective measures of neuropathology, such as nerve conduction velocities; and of specific indicators of dioxin intoxication, such as elevated blood lipids and/or disturbance in porphyrin pigment metabolism.

Finally, it would be incorrect to state that the Vietnamese pilot studies reviewed here have provided definitive evidence for the health effects of phenoxy

---

and so on), and lymphatic structures. However, cancers of the lymphatic system—Hodgkin's disease (Rosenberg, 1982), Burkitt's lymphoma (Ziegler, 1982) and other malignant lymphomas (DeVita *et al.*, 1982)—are not usually subsumed under the term 'soft-tissue sarcoma'.

herbicides. Nevertheless, sufficient evidence has certainly been presented in these studies to warrant additional international attention to the possible health effects of phenoxy herbicides and their contaminants on those people who were exposed during the Second Indochina War: civilians and soldiers from Viet Nam, the USA, Australia and elsewhere.

References

Beale, M. G., Shearer, W. T., Karl, M. M. and Robson, A. M. 1977. Long-term effects of dioxin exposure. *Lancet*, London, **1977**(1): 748.

Carter, C. D. *et al.* 1975. Tetrachlorodibenzodioxin: an accidental poisoning episode in horse arenas. *Science*, Washington, **188**: 738–740.

Creso, E., Marino, V. de, Donatelli, L. and Pagnini, G. 1978. [Neuropsychopharmacological effects of TCDD.] (In Italian) *Bollettino Società Italiana Biologia Sperimentale*, Naples, **54**: 1592–1596.

Danon, J. M., Karpati, G. and Carpenter, S. 1978. Subacute skeletal myopathy induced by 2,4-dichlorophenoxyacetate in rats and guinea pigs. *Muscle and Nerve*, Boston, Mass., **1**(1): 89–102.

DeVita, V. T., Jr, Fisher, R. I., Johnson, R. E. and Berard, C. W. 1982. Non-Hodgkin's lymphomas. In: Holland, J. F. & Frei, E., III (eds). *Cancer Medicine*. Philadelphia: Lea & Febiger, 2465 pp.: pp. 1502–1537.

Drill, V. A. and Hiratzka, T. 1953. Toxicity of 2,4-dichlorophenoxyacetic acid and 2,4,5-trichlorophenoxyacetic acid. *Archives of Industrial Hygiene and Occupational Medicine*, Boston, Mass., **7**: 61–67.

Eberstein, A. and Goodgold, J. 1979. Experimental myotonia induced in denervated muscles by 2,4-D. *Muscle and Nerve*, Boston, Mass., **2**: 364–368.

Elo, H. and Ylitalo, P. 1977. Substantial increase in the levels of chlorophenoxyacetic acids in the CNS of rats as a result of severe intoxication. *Acta Pharmacologica et Toxicologica*, Copenhagen, **41**: 280–284.

Elo, H. A. and Ylitalo, P. 1979. Distribution of 2-methyl-4-chlorophenoxyacetic acid and 2,4-dichlorophenoxyacetic acid in male rats: evidence for the involvement of the central nervous system in their toxicity. *Toxicology and Applied Pharmacology*, New York, **51**: 439–446.

Epstein, S. S., Brown, L. O. and Pope, C. 1982. *Hazardous Waste in America*. San Francisco: Sierra Club Books, 593 pp.

Eriksson, M., Hardell, L., Berg, N. O., Möller, T. and Axelson, O. 1981. Soft-tissue sarcomas and exposure to chemical substances: a case-referent study. *British Journal of Industrial Medicine*, London, **38**: 27–33.

Goldstein, N. P., Jones, P. H. and Brown, J. R. 1959. Peripheral neuropathy after exposure to an ester of dichlorophenoxyacetic acid. *Journal of the American Medical Association*, Chicago, **171**: 1306–1309.

Hardell, L. 1981. Relation of soft-tissue sarcoma, malignant lymphoma and colon cancer to phenoxy acids, chlorophenols and other agents. *Scandinavian Journal of Work, Environmental & Health*, Helsinki, **7**: 119–130.

Hardell, L. and Sandström, A. 1979. Case-control study: soft-tissue sarcomas and exposure to phenoxyacetic acids or chlorophenols. *British Journal of Cancer*, London, **39**: 711–717.

Hardell, L., Eriksson, M., Lenner, P. and Lundgren, E. 1981. Malignant lymphoma and exposure to chemicals, especially organic solvents, chlorophenols and phenoxy acids: a case-control study. *British Journal of Cancer*, London, **43**: 169–176.

Hay, A. 1979. Accidents in trichlorophenol plants: a need for realistic surveys to ascertain risks to health. *Annals of the New York Academy of Sciences*, New York, **320**: 321–324.

Hay, A. 1982. *Chemical Scythe: Lessons of 2,4,5-T and Dioxin*. London: Plenum Press, 275 pp.

Hill, E. V. and Carlisle, H. 1947. Toxicity of 2,4-dichlorophenoxyacetic acid for experimental animals. *Journal of Industrial Hygiene and Toxicology*, Chicago, **29**(2): 85–95.

Honchar, P. A. and Halperin, W. E. 1981. 2,4,5-T, trichlorophenol, and soft tissue sarcoma. *Lancet*, London, **1981**(1): 268–269.

Huff, J. E., Moore, J. A., Saracci, R. and Tomatis, L. 1980. Long-term hazards of polychlorinated dibenzodioxins and polychlorinated dibenzofurans. *Environmental Health Perspectives*, Research Triangle Park, N. Car., **1980**(36): 221–240.

IARC (International Agency for Research on Cancer). 1982. *IARC Monographs on the Evaluation of the Carcinogenic Risk of Chemicals to Humans. Suppl. No. 4.* Lyon: World Health Organization International Agency for Research on Cancer, 292 pp.

JRB Associates. 1981. *Review of Literature on Herbicides, including Phenoxy Herbicides and Associated Dioxins. I. Analysis of Literature. II. Annotated Bibliography.* Washington: US Veterans Administration Department of Medicine & Surgery, 325 + 400 pp.

Kay, J. H., Palazzolo, R. J. and Calandra, J. C. 1965. Subacute dermal toxicity of 2,4-D. *Archives of Environmental Health,* Chicago, **11**: 648–651.

Kimbrough, R. D., Carter, C. D., Liddle, J. A., Cline, R. E. and Phillips, P. E. 1977. Epidemiology and pathology of a tetrachlorodibenzodioxin poisoning episode. *Archives of Environmental Health,* Chicago, **32**: 77–86.

Kociba, R. J. *et al.* 1978. Results of a two-year chronic toxicity and oncogenicity study of 2,3,7,8-tetrachlorodibenzo-*p*-dioxin in rats. *Toxicology and Applied Pharmacology,* New York, **46**: 279–303.

Kouri, R. E. *et al.* 1978. 2,3,7,8-tetrachlorodibenzo-*p*-dioxin as a cocarcinogen causing 3-methylcholanthrene-initiated subcutaneous tumors in mice genetically "nonresponsive" to Ah locus. *Cancer Research,* Philadelphia, **38**: 2777–2783.

Morton, D. L. and Eilber, F. R. 1982. Soft tissue sarcomas. In: Holland, J. F. & Frei, E., III (eds). *Cancer Medicine.* Philadelphia: Lea & Febiger, 2465 pp.: pp. 2141–2157.

Oliver, R. M. 1975. Toxic effects of 2,3,7,8 tetrachlorodibenzo 1,4 dioxin in laboratory workers. *British Journal of Industrial Medicine,* London, **32**: 49–53.

Palmer, J. S. 1963. Chronic toxicity of 2,4-D alkanolamine salts to cattle. *Journal of the American Veterinary Medical Association,* Schaumburg, Ill., **143**: 398–399.

Palmer, R. S. and Radeleff, R. D. 1964. Toxicologic effects of certain fungicides and herbicides on sheep and cattle. *Annals of the New York Academy of Sciences,* New York, **111**: 729–736.

Pitot, H. C., Goldsworthy, T., Campbell, H. A. and Poland, A. 1980. Quantitative evaluation of the promotion by 2,3,7,8-tetrachlorodibenzo-*p*-dioxin of hepatocarcinogenesis from diethylnitrosamine. *Cancer Research,* Philadelphia, **40**: 3616–3620.

Pocchiari, F., Silano, V. and Zampieri, A. 1979. Human health effects from accidental release of tetrachlorodibenzo-*p*-dioxin (TCDD) at Seveso, Italy. *Annals of the New York Academy of Sciences,* New York, **320**: 311–320.

Rosenberg, S. A. 1982. Hodgkin's disease. In: Holland, J. F. & Frei, E., III (eds). *Cancer Medicine.* Philadelphia: Lea & Febiger, 2465 pp.: pp. 1478–1502.

Singer, R., Moses, M., Valciukas, J., Lilis, R. and Selikoff, I. J. 1982. Nerve conduction velocity studies of workers employed in the manufacture of phenoxy herbicides. *Environmental Research,* New York, **29**: 297–311.

Tung, Ton That. 1973. [Primary cancer of the liver in Viet Nam.] (In French) *Chirurgie,* Paris, **99**: 427–436.

Van Miller, J. P., Lalich, J. J. and Allen, J. R. 1977. Increased incidence of neoplasms in rats exposed to low levels of 2,3,7,8-tetrachlorodibenzo-*p*-dioxin. *Chemosphere,* Elmsford, N.Y., **6**(9): 537–544 [also **6**(10): 625–632].

Zack, J. A. and Suskind, R. R. 1980. Mortality experience of workers exposed to tetrachlorodibenzodioxin in a trichlorophenol process accident. *Journal of Occupational Medicine,* Downers Grove, Ill., **22**(1): 11–14.

Ziegler, J. L. 1982. Burkitt's lymphoma. In: Holland, J. F. and Frei, E., III (eds). *Cancer Medicine.* Philadelphia: Lea & Febiger, 2465 pp.: pp. 1537–1546.

Chapter Seven
Reproductive Epidemiology

7.A. Reproductive epidemiology: Symposium summary[1]

Nguyên Cân et al.[2]

Institute for the Protection of Mother and Newborn, Hanoi

This summary report examines epidemiologically the long-term effects on human reproduction of the herbicides applied in South Viet Nam during the Second Indochina War of 1961–1975. Of particular concern here is the dioxin contaminant of Agent Orange, the major herbicide that was employed.

The Working Group accepts without dissent the animal evidence proving the teratogenicity (causing of birth defects) of dioxin when administered to females, but remains unaware of any acceptable evidence of the transmission of this toxicity through the male.

Although there have been many studies of the medical effects of Agent Orange and related compounds, together with their contaminants, these studies have been inconclusive as regards reproductive effects. Therefore, the study of these effects in Viet Nam, where there has been such extensive exposure, seems to be of the greatest interest and importance, not only to Viet Nam but also to the rest of the world.

Recognizing—and, indeed, deeply cognizant of—the extraordinary difficulties necessarily associated with any such retrospective study, especially when it is being carried out more than a decade after the time of exposure, the Working Group has been very much impressed by the seven Vietnamese studies that have been reported to it. These Vietnamese evaluations of the possible teratogenic and/or mutagenic (causing genetic, i.e., DNA or gene, damage) effects of herbicide exposure are being made in the following three major ways: (a) changes in the frequency of spontaneous abortions (miscarriages) and still births relative to normal deliveries; (b) changes in the frequency of congenital malformations (birth defects); and (c) changes in the rate of occurrence of molar pregnancies (where a hydatidiform mole develops in lieu of a foetus).

[1] Summary report of a Working Group of the International Symposium on Herbicides and Defoliants in War: the Long-term Effects on Man and Nature, Ho Chi Minh City, 13–20 January 1983 (appendix 3).
[2] The Working Group consisted of: Trinh Kim Anh, L. Bisanti, Nguyên Cân (Chairman), Nguyên Trân Chiên, J. D. Constable (Rapporteur), Z. Fülöp, M. C. Hatch, Bui Sy Hung, Lê Diêm Huong, Nguyên Dình Khoa, M. Kida, J. Kucera, Tôn Dúc Lang, Nguyên Xuan Loc, Hô Dang Nguyên, Om Sokha, Nguyên Thi Ngoc Phuong (Vice-chairwoman), V. Rajpho, Sau Sok Khonn, T. D. Sterling and Nguyên Thi Xiem.

Changes in spontaneous abortions, still births and congenital anomalies have been studied not only among exposed women (necessarily in southern Viet Nam), but also in the children of *un*exposed women whose husbands had been exposed.

The authors of all these investigations, well aware of the many obstacles to a completely satisfactory study, have proffered them as preliminary reports even when they already include an immense amount of laboriously acquired information.

The most complete, and perhaps consequently the most impressive and persuasive, studies relate to an increase in the unfavourable outcomes of pregnancy in North Vietnamese women whose husbands served in the South during the war, and were therefore at least potentially exposed to herbicides, as compared to fellow villagers whose husbands had remained in the North. Providing that the following criteria have, indeed, been met—a properly carried out blind study; an absence of bias, especially in selection; maximum validation of data other than by self-reporting; and strict adherence to a properly prepared protocol (and the Working Group has no specific reason to doubt any of them)— then a statistically valid increase in these unfavourable outcomes has been shown for the unexposed wives of exposed fathers in one study (Cân, paper 7.B), and is strongly suggested in another which additionally indicates a reversal of the usual increase in the frequency of such disasters with progressive pregnancies.

However, it is agreed by all that in such investigations—especially when they show results contrary to previous experimental evidence—one, or even two or three, congruent investigations are not enough to provide complete proof of their conclusions, and further similar work is needed.

As regards congenital anomalies, there are several studies apparently indicative of a generally higher rate of their frequency among exposed women, but these changes are often hard to prove beyond any doubt. The absolute rate of reported congenital anomalies in Viet Nam seems generally very low. Although the reasons for this are not fully understood, they may include the low sensitivity of the information system, a reduced exposure to toxic chemicals, and inherent ethnic differences.

The Working Group is much impressed by the large numbers of reported cases of the following five categories of congenital anomalies: (*a*) anencephaly (absence of the brain) and other neural tube defects, which here are associated with a remarkably low incidence of spina bifida (defective closure of the spinal cord); (*b*) deformities of the sensory organs, such as anophthalmia (eyes missing); (*c*) deformities of the limbs, including phocomelia; (*d*) conjoined twins; and (*e*) orofacial cleft defects. In most other countries these malformations are either not very common (anencephaly, orofacial clefts) or even rare (deformities of the limbs, deformities of the sensory organs, conjoined twins).

In order to pursue further this important field of inquiry appropriately, at least the following two things are essential: (*a*) a more precise identification and classification of reported anomalies; and (*b*) a determination of the expected rate of such deformities in Viet Nam. The latter can be accomplished in one of three ways: (*a*) best, by the recovery of accurate pre-spray figures; (*b*) second best, by the use of data from more or less closely related populations, available through the World Health Organization (WHO) for the years 1962–1963 for India, Hong Kong, Malaysia and Singapore; or (c) minimally desirable, by determination of the world-wide reported ranges of frequency.

Molar pregnancies seem to show an increase in frequency in herbicide-exposed women, but more work is needed if this is to be proved. Recognizing the relatively high frequency of this lesion in South-east Asia, information should be obtained as to any recent changes reported for elsewhere in the area.

The Working Group feels that the general design of the studies reported is excellent, but that additional numbers are needed, that controls be made somewhat stricter, that possible variables be carefully scrutinized, and that the protocols be rigidly adhered to. Until now, exposure index has rarely been included and no sort of dose–response curve has been constructed. Consideration should be given to comparing the possible results of direct exposure compared to exposure via the diet. A search should perhaps be made for other possible toxins such as heavy metals or DDT.

Finally the Working Group would point out that even if all of these studies, as designed, were to yield unequivocally positive results, then only the increased defects resulting from exposure would have been proved, not a specific association with dioxin. The latter would remain presumptive until the causal relationship was confirmed by separate investigations. These would be made easier if the newest methods of chemical analysis can, indeed, still demonstrate residual dioxin at variable levels in human tissue.

The Working Group agrees that the remaining problems of the possible teratogenesis of herbicides require extensive continued study by the scientists of Viet Nam in which they could be appropriately aided by the international scientific community, especially with respect to experimental laboratory investigations.

7.B. Effects on offspring of paternal exposure to herbicides

Nguyên Cân[1]

Institute for the Protection of Mother and Newborn, Hanoi

I. Introduction

During the Second Indochina War large areas of farm and forest were sprayed with herbicides for hostile purposes, especially from 1966 to 1969. Many people living or serving there during that period were inevitably exposed to these chemicals, either directly or indirectly. The major herbicide employed was Agent Orange (together with its dioxin impurity), accounting for more than 80 per cent of the total in terms of weight of active ingredients.

In the present study the question asked was whether wartime exposure by males had an effect on the offspring they conceived during that time with unexposed females.

II. Materials and methods

The study area was three agricultural (rice-growing) districts in North Viet Nam: (*a*) Mai Chau district, Ha Son Binh province, a mountainous region about 85 km west-south-west of Hanoi; (*b*) My Van district, Hai Hung province, a plains region about 50 km east-south-east of Hanoi; and (*c*) Hai Hau district, Ha Nam Ninh province, a coastal region about 100 km south-south-east of Hanoi. A total of 40 064 local women were selected for inclusion in the study (of which roughly 14 per cent were from the first district, 40 per cent from the second, and the remaining 46 per cent from the third). All of the women had been married and pregnant one or more times during the war (an average of 3.8 pregnancies each); none were recorded as having had a history of tuberculosis, syphilis or malaria; and apparently none had taken antibiotics or hormones during their pregnancies. Each woman was interviewed in detail during June–August 1982 by a physician or midwife (the latter under the direction of a physician) regarding her health history and wartime pregnancies; and the information thus gained was augmented by examination of the District health

---

[1] Adapted by the editor from the author's Symposium presentation.

records. It was established in particular whether and when the woman's husband had served in South Viet Nam during the war (thus presumably having been exposed to the herbicides sprayed). In about one-fifth of the pregnancies studied the husband had been serving in the South prior to conception; and in the remaining four-fifths he had remained in the unsprayed North.

III. Results

The data suggest that when women living in North Viet Nam during the war (and thus not exposed to herbicides) became pregnant by men who had been serving in South Viet Nam during that period (and thus presumably exposed to herbicides) their pregnancies were slightly more likely to end in spontaneous abortion, although (perhaps as a result) less likely to come to term as still births, than if the father had not been to the South (table 7.B.1). Moreover, among the resulting live births there appears to have been a somewhat greater frequency of congenital malformations (birth defects). The approximately 30 per cent reduction in known pregnancies among women with husbands serving in war-time South Viet Nam might have found its basis in an increased frequency of early, undetected spontaneous abortions, but may also have been the result of wartime separation.

To what extent the observed increase in the frequency of congenital defects among the offspring of fathers who had served in wartime South Viet Nam was significant was tested not only by a 'chi-square' test (table 7.B.1), but also by a case-control (case-referent) approach. Sixty-one families with a surviving

Table 7.B.1. **Wartime pregnancies in North Viet Nam in relation to father's service in South Viet Nam (and presumed herbicide exposure)**

Outcome of pregnancy	Neither parent exposed[a] (No./1 000)	Father exposed[b] (No./1 000)	Change[c] (per cent)
Normal birth	910	898	−1
Congenital defect[d]	4.27	5.89	+38
Still birth	20.6	18.0	−13
Spontaneous abortion	58.6	70.8	+21
Curettage	6.15	6.55	+7
Molar pregnancy	0.574	0.873	+52
Total	**1 000**	**1 000**	

[a] Based on 121 993 pregnancies among 29 041 women, that is, 4.2 pregnancies per woman (annex 7.B.1).
[b] Based on 32 069 pregnancies among 11 023 women, that is, 2.9 pregnancies per woman (annex 7.B.1).
[c] A 'chi-square' test by the author suggests that none of the differences is significant at the 1 per cent level, but that the increase in spontaneous abortions is significant at the 5 per cent level and that of congenital defects at the 10 per cent level.
[d] Of which perhaps one-tenth are fatal.

congenitally defective child were chosen at random from among those studied above, and each of these was matched for age, number of deliveries, living environment, and age of offspring with three families with normal children. It turned out that among the 61 test families, the fathers in 30 had served in the South, that is, 49 per cent of them. In contrast, among the 183 control families it turned out that only 39 had, that is, 21 per cent. These results would suggest that the father of a congenitally defective child was more than twice as likely to have served in the South than the father of a normal child, a difference that was found to be significant at the 5 per cent level.

In addition to a possible slight overall increase in congenital defects among pregnancies resulting from males who had served in the South, there appears to have been an increase within this category in the proportion of cleft lips and perhaps of anencephaly (absence of the brain), together with a decrease in the proportion of spina bifida (defective closure of the spinal cord) and of distended belly.

IV. Conclusion

The tentative conclusion is reached that the pregnancies resulting from males who had been exposed to the wartime herbicides, and mating with unexposed females, were somewhat more likely to end in spontaneous abortion; and that those pregnancies that came to term as live births were somewhat more likely to suffer from congenital defects, especially cleft lip. Such an effect on offspring via exposure of the father appears not to have been previously described and thus deserves rigorous follow-up studies.

Annex 7.B.1. Wartime pregnancies in North Viet Nam in relation to father's service in South Viet Nam (and presumed herbicide exposure): summary of the numerical data

Outcome of pregnancy	Mountainous region[a]		Plains region[b]		Coastal region[c]		Combined sample[d]	
	Neither parent exposed	Father exposed	Neither parent exposed	Father exposed	Neither parent exposed	Father exposed	Neither parent exposed	Father exposed
Normal birth	16 452	1 773	42 002	13 807	52 538	13 215	110 992	28 795
Congenital defect[e]	118	28	195	103	208	58	521	189
Still birth	663	34	820	289	1 029	253	2 512	576
Spontaneous abortion	1 680	189	2 651	995	2 817	1 087	7 148	2 271
Curettage	297	39	271	85	182	86	750	210
Molar pregnancy	9	2	43	10	18	16	70	28
Total	**19 219**	**2 065**	**45 982**	**15 289**	**56 792**	**14 715**	**121 993**	**32 069**

[a] Mai Chau district, Ha Son Binh province.
[b] My Van district, Hai Hung province.
[c] Hai Han district, Ha Nam Ninh province.
[d] A total of 40 064 unexposed women were involved in the combined sample, 29 041 of them impregnated by unexposed males and the remaining 11 023 by exposed males.
[e] The frequency of congenital defects appears low in comparison with values for other countries, but this appears to stem from the restriction of the above data to grossly detectable defects (and not, e.g., including cardiac or blood vessel defects). Of the order of one-tenth of these defects are fatal.

7.C. Reproductive epidemiology: an overview[1]

Arthur H. Westing

Stockholm International Peace Research Institute

I. Introduction

In any consideration of the possible medical effects of exposure to herbicides, great interest attaches to the possibility that they cause an increase in adverse outcomes of pregnancy, that is, in: (*a*) spontaneous abortions (miscarriages); (*b*) still births; (*c*) molar pregnancies (in which a hydatidiform mole develops in lieu of a foetus); and (*d*) offspring with congenital malformations (birth defects).

Of the herbicides sprayed for hostile purposes during the Second Indochina War, Agent Orange stands out because of its composition—half 2,4,5-T (with its dioxin contaminant) and half 2,4-D—and because of the magnitude of its use—more than 60 per cent of what was sprayed in terms of volume (and more than 80 per cent in terms of active ingredients). Animal experimentation suggests that dioxin especially can lead to adverse outcomes of pregnancy (Hay, paper 8.C).

First to be discussed in the present overview are some of the problems inherent in epidemiological studies of human reproduction, followed by a brief review of the literature. Next summarized and evaluated are some recent Vietnamese studies. The paper concludes with some recommendations for future research.

II. Problems in reproductive epidemiological research

That dioxin causes reproductive anomalies in a number of small laboratory mammals is now undisputed. However, known variation in species sensitivity as well as differences in sexual habits make it impossible to look for the information we need here other than in humans.

Reproductive function in humans can be affected adversely by an agent in one or more of several ways depending on the nature and timing of the exposure and whether it acts upon the mother, the father, or the foetus. The relevant range of effects includes male sterility, female sterility, spontaneous abortion,

[1] This paper was prepared by the editor based on information supplied by John D. Constable, Maureen C. Hatch and other members of the Symposium Working Group.

still birth and congenital malformation. When the mother is exposed, there is considered to be a potential for damage regardless of how long prior to conception the exposure occurs, in that the female germ cells are all present at the time the mother was born. An exposure that occurs during pregnancy could act on the foetus directly or it could act indirectly by altering the maternal milieu. (Effects on the foetus resulting from exposure during pregnancy, especially those resulting in congenital defects, are generally labelled as teratogenic effects.)

When it is the father that has been exposed, the range of effects that may follow includes: reduction in libido; reduction in potency, perhaps to the point of infertility; or damage, including genetic damage, to the germ cells. It is important to point out that the possibility that exposure of a male to a chemical agent might lead to teratogenic damage to his offspring has apparently not been described for humans; it has, however, been reported for a laboratory animal for one substance, not an herbicide (Lutwak-Mann, 1964). In considering paternal exposure in humans, the period of maximum vulnerability might be thought to be at or shortly before conception. For the damage to occur to the sperm itself the period could not extend beyond about 80 d; however, if it is the sperm precursor cells that are vulnerable, there would be no such strict time limitation. Post-conception paternal influence is also conceivable (though undemonstrated) via post-conception intercourse if a toxic agent were to concentrate in the semen.

There are problems inherent in epidemiological investigations beyond those peculiar to reproductive functioning. Studies carried out in human populations are fraught with all the difficulties of uncontrolled research. Inevitably differences will exist between exposed and unexposed subjects that allow for alternative explanations for observed effects. Even in developed countries, with detailed reliable demographic and public health statistics, such studies require large samples, long periods of time, a sound theoretical framework, and sophisticated design and analysis.

III. Review of the literature

The studies where human exposure to phenoxy herbicides or their dioxin contaminants has been examined in relation to reproductive function are few in number, highly varied in their details and to a large extent non-definitive. Thus, two studies can be cited of occupational exposure in neither of which the sample was sufficiently large or well defined to produce valid evidence regarding congenital malformations. In the first of these, no excess of reproductive problems was reported among the offspring of male factory workers occupationally exposed to dioxin (and whose wives were unexposed) with the possible exception of a slight increase in the spontaneous abortion rate in one subset of the sample (Townsend et al., 1982). Similarly, in the second study there was no reported increase in reproductive anomalies among the offspring of male 2,4,5-T spray applicators (Smith et al., 1982).

Of the several studies in which an attempt has been made to link environmental exposure to phenoxy herbicides sprayed for agricultural or forestry purposes with reproductive anomalies among the local populace, none provides satisfactorily interpretable data (Field & Kerr, 1979; Hanify *et al.*, 1981; Nelson *et al.*, 1979; Thomas, 1980; Thomas & Czeizel, 1982). Studies of this kind (i.e., those carried out at the group rather than individual level and relying on aggregate data) are so susceptible to problems of misclassification and to the effects of confounding variables that results may overestimate, underestimate, or even reverse the true relationship between exposure and outcome (Robinson, 1950). Regarding accidental exposure, the possibility exists that there has been some increase in spontaneous abortions, still births and congenital defects among the population that was exposed to dioxin following the factory explosion in Sèveso, Italy in 1976, but the numbers here are too small to be reliable (Bisanti *et al.*, 1980; Reggiani, 1979).

A number of studies have emerged from the Second Indochina War, of which the earlier ones are noted here and the several recent Vietnamese studies reviewed in the section following this one. Cutting *et al.* (1970) compared the frequency of reproductive anomalies (still births, molar pregnancies, congenital defects) in selected hospitals in South Viet Nam for the no-spray/light-spray period 1960–65 with the heavy-spray period 1966–69. Although there were some indications of increased rates, this study is seriously flawed because of inadequacies in the data base (see below).

Kunstadter (1982) during the 1970s studied the frequencies of still births and congenital defects in South Viet Nam on the basis of hospital records and maternal interviews, together with relevant spray information from US Air Force records. He reported an overall decrease in congenital defects during the spray period, but a proportional increase in cleft lip during the period of heavy spraying. He further noted an increase in still births following the spray period. The methodological limitations of this study make its definitive interpretation difficult.

Donovan *et al.* (1983) reported on a case-control (case-referent) study in Australia involving 8 517 cases of congenital defects. No association could be discerned between service by the father in Viet Nam and congenital defects in his offspring. However, "exposure to herbicides was infrequent and probably very low in Australian troops in Vietnam" (page 2) and thus this study cannot be considered significant in evaluating evidence on the reproductive effects of herbicide exposure.

In summary, no study published to date seems to be conclusive in either proving or disproving an association of phenoxy herbicide/dioxin exposure with adverse outcomes of pregnancy in humans (see also Hatch & Kline, 1981).

IV. Recent Vietnamese studies

A number of recent unpublished studies are summarized and evaluated here, several carried out in southern Viet Nam and the remainder in northern Viet Nam.

Nguyên Dình Khoa (University of Hanoi, unpublished) recently obtained medical histories by interview for the 17-year period 1965–82 in several hill-tribe (Montagnard) villages in A Lưới district, Bình Trị Thien province (in the former Thua Thien province), roughly 100 km west of Danang, two in an area of heavy wartime herbicide spraying and the other in an unsprayed area. For the sprayed area Khoa's data suggested a spontaneous abortion rate of 10 per cent (134 abortions of 1 330 pregnancies) and for the unsprayed area 6 per cent (38 abortions of 625 pregnancies). Among those pregnancies that came to term in one of the sprayed villages, approximately 3 per cent resulted in congenital defects (apparently 23 congenital defects of 820 pregnancies; or of 744 deliveries); in contrast, in the unsprayed village, the rate was approximately 1 per cent (apparently 6 congenital defects of 625 pregnancies; or of 587 deliveries). Unfortunately, Khoa's methodology, data and results do not lend themselves to analysis or evaluation.

Hô Dang Nguyên (Provincial Polyclinic Hospital, Tay Ninh City, unpublished) has examined the records of reproductive outcome during the period from 1979 to mid-1982 at the major hospital in Tay Ninh province, a province that had been subjected to heavy wartime herbicide spraying. Among a group of 57 congenital defects referred to the hospital during that period (and hence a selected sample), only one mother had a history of malaria, two tested positive for syphilis and very few smoked or had used medications during their pregnancy. Among these 57 defects, 2 were so-called monsters, 8 showed anencephaly (absence of the brain), 4 anophthalmia (absence of the eyes), 4 phocomelia (shortened limbs) and 4 others severe skeletal deformities. Although not susceptible to statistical analysis, this record of congenital defects can be considered to represent a larger than expected number of striking (i.e., rare) cases.

Cung Bính Trung & Nguyên Trân Chiên (College of Medicine, Hanoi, unpublished) have recently studied rates of spontaneous abortion and congenital defects before and after the period of spraying in both sprayed and unsprayed communes, among a total of 848 families (method of selection not indicated). The study area was within Giong Trom district, Ben Tre province (in the former Kien Hoa province), approximately 75 km south-south-west of Ho Chi Minh City. Among those living in sprayed communes the pre-spray-period spontaneous abortion rate was 6 per cent (129 abortions of 2 292 pregnancies); their post-spray-period spontaneous abortion rate rose to 14 per cent (217 abortions of 1 562 pregnancies). Among those in presumably comparable families living in unsprayed villages, the pre-spray-period spontaneous abortion rate was 7 per cent (32 abortions of 436 pregnancies) and the post-spray-period spontaneous abortion rate remained essentially unchanged, also at 7 per cent (27 abortions of 365 pregnancies). To what extent the increased age—known to result in more frequent birth defects—had an effect on these rates is not clear. Trung & Chiên also studied congenital defects in the sprayed (but not in the unsprayed) communes, apparently excluding cases induced by such factors as syphilis or drugs. Prior to the spray period these represented 1.3 cases per thousand pregnancies (3 congenital defects of 2 292 pregnancies; or 1.4 cases per thousand surviving births, i.e., 3 of 2 163). Following the spray period these represented 15.4 cases per thousand pregnancies (24 congenital defects of 1 562 pregnancies; or 17.8 cases per thousand surviving births, i.e., 24 of 1 345). The pre-exposure rate seems to reflect very limited reporting; and again, mothers were older in the post-exposure period.

Lê Diêm Huong & Nguyên Thi Ngoc Phuong (Tu Du Nhi Dong Hospital, Ho Chi Minh City, unpublished) have recently compiled pregnancy data for selected years during the period 1952–1981 (table 7.C.1). Using data from their institution (a major gynaeco-obstetrical hospital), they compared frequencies of a number of adverse outcomes of pregnancy before, during (primarily 1966– 1969) and since the herbicide spraying in South Viet Nam. Their data suggest a dramatic increase in the rate of spontaneous abortion following the onset of heavy spraying, which appears only now to be tapering off. However, in view of the remarkably low pre-spray values, the reported increase may well have been the result of improved ascertainment rather than an effect of exposure. The data

Table 7.C.1. Pregnancies at Tu Du Nhi Dong Hospital, Ho Chi Minh City, 1952–1981 (selected years)[a]

Year[b]	Number of pregnancies	Spontaneous abortions (No./1 000)	Still births[c] (No./1 000)	Molar pregnancies plus uterine cancers[d] (No./1 000)	Gross congenital defects[e] (No./1 000)
1952	6 495	4	6	9	
1953	6 889	15	1		
1959	14 413			12	
1960	14 076			11	
1962	12 440			13	
1963	12 538				7
1964	16 779				6
1965	19 308	47			5
1966	19 854	41			4
1967	24 345	148	16	14	5
1971	27 475	139	14	9	
1976	16 167	169	13	21	
1977	12 943	165	18	39	7
1978	13 544	181	14	35	8
1979	12 757	109	15	45	11
1980	12 766	127	14	36	13
1981	13 574	101	18	45	10

[a] Derived by the editor from data compiled by Lê Diêm Huong & Nguyên Thi Ngoc Phuong (Tu Du Nhi Dong Hospital, Ho Chi Minh City, unpublished). The data for the years 1952–1971 are primarily from unpublished doctoral theses; those for the years 1976–1981 are from the annual reports of the Hospital. The missing data are unavailable.
[b] The period of major herbicide spraying in South Viet Nam was 1966–1969.
[c] These were referred to as intra-uterine deaths.
[d] Molar pregnancies (where a hydatidiform mole develops in lieu of a foetus). The uterine cancers referred to here are choriocarcinomas (cancers of the membrane surrounding the foetus).
[e] This enumeration includes only gross congenital defects such as a cleft lip, the absence of a limb, anencephaly (absence of the brain) and hydrocephaly (enlarged head with atrophy of the brain and so on).

further suggest that molar pregnancies plus certain uterine cancers (chorio-carcinomas) have experienced a sharp increase since 1976. The data for still births and gross congenital defects are insufficiently complete to permit the identification of any possible changes in rate through time. The interpretation of the data of Huong & Phuong is also made difficult by the change in the source of the data mid-way through the series (see table 7.C.1, note *a*).

As a means of relating the increased incidence of molar pregnancies to wartime herbicide exposure, Huong & Phuong (unpublished) carried out a small case-control study in which 85 women with molar pregnancies were matched with 276 women with normal deliveries. (This study had begun with 100 test cases plus 284 matched controls; no information was provided on the excluded 15 cases plus 8 controls and no reason given for their exclusion.) The members of the control group were well matched with those of the test group for age, but less well for social conditions, diet, and habits of the husbands. Interviews disclosed that 48 of the 85 test women had been exposed, that is, 56 per cent of them. In contrast, only 27 of the 276 control women had been exposed, that is, 10 per cent of them.

The data of Huong & Phuong on molar pregnancies are impressive, suggesting a strong association with herbicide exposure. Determination of dioxin levels in the mole or subsequent choriocarcinoma, or in the adipose (fat) or other tissues, of the patient would be very helpful in confirming this possible association.

Nguyên Thi Ngoc Phuong & Lê Diêm Huong (unpublished) have recently compared reproductive problems of a herbicide-exposed group with an un-exposed group (table 7.C.2). They found higher frequencies of spontaneous abortions, still births, molar pregnancies and congenital defects among the exposed group. No time period for the reproductive histories is given nor is the method for selecting the subjects, making interpretation of these data proble-matical.

Tôn Dúc Lang & Dô Dúc Vân (Viet Duc Huu Nghi Hospital, Hanoi, unpublished) have carried forth preliminary investigations by Ton That Tung of delayed adverse outcomes of pregnancy resulting from herbicide exposure

Table 7.C.2. Pregnancies in South Viet Nam in relation to parental herbicide exposure[a]

Outcome of pregnancy	Parents unexposed[b] (No./1 000)	Parents exposed[c] (No./1 000)
Spontaneous abortion	36	80
Still birth	0.30	8.1
Molar pregnancy	3.9	18
Congenital defect	4.3	11

[a] Derived by the editor from data recently compiled by Nguyên Thi Ngoc Phuong & Lê Diêm Huong (Tu Du Nhi Dong Hospital, Ho Chi Minh City, unpublished). The time period covered is not indicated.
[b] Based on 6 690 pregnancies among 1 126 families (i.e., 5.9 pregnancies per woman) in District No. 10, Ho Chi Minh City.
[c] Based on 7 327 pregnancies among 1 249 families (i.e., 5.9 pregnancies per woman) in Thanh Phong village, Thanh Phu district, Ben Tre province.

of the father. In one instance they investigated the frequency of congenital defects among the offspring of ex-soldiers in the three immediate post-war year (1975–1978) on the basis of the hospital records in Yen Bai district, Hoang Lien Son province, about 130 km north-west of Hanoi. Among those families where the father had remained in unsprayed North Viet Nam during the war, the rate of congenital defects was found to be 6 per thousand deliveries (15 of 2 547). In contrast, when the father had served in South Viet Nam during the war (and was thus presumably exposed), this rate was reported to be fully 29 per thousand (15 of 511). Comparable results were obtained in a similar study of a number of unspecified agricultural and handicraft co-operatives in the North: 5 congenital defects per thousand deliveries (10 of 2 172) for unexposed parents versus 23 or 26 per thousand (71 or 82 of 3 147) where the father had served in the South. In these two studies the wives of unexposed fathers show the increase in the frequency of adverse outcomes with successive pregnancies normally expected, whereas this trend is reversed among the women whose husbands were presumably exposed. The pattern is suggestive of a toxic effect most virulent at the time of first conception and then gradually diminishing in its potency. This very interesting report by Lang & Vân that paternal exposure to a herbicide might result in subsequent (delayed) reproductive anomalies still requires independent verification.

In a recent study by Cân (paper 7.B) the more biologically plausible proposition was examined as to whether paternal exposure to herbicides might have an effect on reproductive anomalies during the period of exposure. He concluded that such exposure led to somewhat higher incidences of spontaneous abortions and of congenital defects (especially of cleft lip). The protocol of Cân's study is rather more convincing than that of Lang & Vân's.

In summary, the several recent Vietnamese studies in which both parents were at risk of herbicide exposure are consistent in reporting increases in spontaneous abortions, still births, molar pregnancies and congenital defects. The evidence for an increase in molar pregnancies is very suggestive. The studies in which presumptive paternal exposure was associated with reproductive anomalies are consistent in reporting increases in congenital defects, particularly cleft lip and perhaps anencephaly. The data on spontaneous abortions are conflicting. Moreover, no association between paternal exposure and molar pregnancy is demonstrated.

With respect to the Vietnamese studies in general it may be well to consider a number of potential pitfalls:

1. Bias is a threat to the validity of any scientific research, particularly so when the investigation does not take place in the pristine environment of the laboratory, but rather in the real world. Hence it is unfortunate that some of the Vietnamese investigations prevent adequate evaluation of this matter because they are less detailed than one would wish in describing the selection of subjects and the method of data gathering.

2. The establishment of herbicide exposure in some of the Vietnamese investigations is based merely on residence in an area that had been sprayed.

Obviously there must have been enormous variation in individual exposure: some people were outdoors at the time of spraying, others indoors, and still others perhaps away at the time. Conversely, people from unsprayed areas may have visited sprayed areas. Moreover, no distinction has been drawn in the Vietnamese studies between direct exposure and secondary exposure through the diet. In no instance have the various herbicidal agents been differentiated.

3. The sources of data on reproductive outcomes in some of the Vietnamese investigations have been verbal reports (interviews), which may not be entirely reliable. Other data have been obtained from local health or hospital records which (a) appear to not report about one-quarter of the births and (b) are not fully reliable, especially so with respect to congenital defects (which are not uniformly classified and which appear to be consistently under-reported). For example, Meselson et al. (1972) had occasion to study original midwife birth records in South Viet Nam during 1966–1972 and found the relation between these and the published Ministry of Health figures based on them to be very poor. As a specific example, the Ministry of Health figures for Tay Ninh province which were used by Cutting et al. (1970) are at striking variance with the original local midwife records, seriously flawing this study, as suggested earlier (see also Meselson et al., 1972; Westing, 1978, pages 290–291).

4. Only limited efforts have been made in some of the Vietnamese investigations to control for potentially confounding variables (e.g., maternal age, nutrition, infection) which might distort any observed association between herbicide exposure and reproductive outcome.

V. Conclusion

There is sufficient apparent evidence of adverse reproductive effects in Viet Nam following exposure to herbicides to indicate the need for continued and expanded investigation. The general designs of the various studies now in train there are appropriate, as are the study populations and outcomes of pregnancy being examined. To increase the validity of these inquiries, some methodological improvements are desirable, as has been alluded to earlier. One must perpetually keep in mind that no degree of care in statistical analysis can compensate for erroneous information and may indeed shroud it in verisimilitude. The problem of establishing exposure (including level of exposure) might be enormously facilitated if the technology of measuring residual dioxin in body tissues as well as in the environment could be sufficiently refined. If such tests become available they should be recommended for studies of victims of molar pregnancies and of uterine cancers (especially choriocarcinomas).

The reproductive anomalies reported to result from wartime herbicide exposure are not especially amenable to preventive or remedial action. An exception is the surveillance of women with molar pregnancies for the early detection of choriocarcinoma. The frequency of congenital defects, even when viewed in the most pessimistic light, is certainly not such as to warrant recommending either widespread examination of foetuses for such anomalies (via amniocentesis) or elective abortion.

References

Bisanti, L. *et al.* 1980. Experiences from the accident of Seveso. *Acta Morphologica Academiae Scientiarum Hungaricae*, Budapest, **28**: 139–157.

Cutting, R. T., Phuoc, T. H., Ballo, J. M., Benenson, M. W. and Evans, C. H. 1970. *Congenital Malformations, Hydatidiform Moles, and Stillbirths in the Republic of Vietnam, 1960–1969*. Washington: US Dept of Defense, 29 pp.

Donovan, J. W., Adena, M. A., Rose, G. and Battistutta, D. 1983. *Case-Control Study of Congenital Anomalies and Vietnam Service: Birth Defects Study*. Canberra: Australian Government Publishing Service, 127 pp.

Field, B. and Kerr, C. 1979. Herbicide use and incidence of neural-tube defects. *Lancet*, London, **1979**(1): 1341–1342.

Hanify, J. A., Metcalf, P., Nobbs, C. L. and Worsley, K. J. 1981. Aerial spraying of 2,4,5-T and human birth malformations: an epidemiological investigation. *Science*, Washington, **212**: 349–351.

Hatch, M. and Kline, J. 1981. *Spontaneous Abortion and Exposure during Pregnancy to the Herbicide 2,4,5-T*. Washington: US Environmental Protection Agency Publ. No. EPA 560/6-81-006, 62 pp.

Kunstadter, P. 1982. *Study of Herbicides and Birth Defects in the Republic of Vietnam: An Analysis of Hospital Records*. Washington: National Academy Press, 73 pp.

Lutwak-Mann, C. 1964. Observations on progeny of thalidomide-treated male rabbits. *British Medical Journal*, London, **1964**: 1090–1091.

Meselson, M. S., Westing, A. H. and Constable, J. D. 1972. Background material relevant to presentations at the 1970 annual meeting of the AAAS. *US Congressional Record*, Washington, **118**: 6807–6813.

Nelson, C. J., Holson, J. F., Green, H. G. and Gaylor, D. W. 1979. Retrospective study of the relationship between agricultural use of 2,4,5-T and cleft palate occurrence in Arkansas. *Teratology*, Philadelphia, **19**: 377–383.

Reggiani, G. 1979. Estimation of the TCDD toxic potential in the light of the Seveso accident. *Archives of Toxicology*, W. Berlin, **1979**(Suppl. 2): 291–302.

Robinson, W. S. 1950. Ecological correlations and the behavior of individuals. *American Sociological Review*, Washington, **15**: 351–357.

Smith, A. H., Fisher, D. O., Pearce, N. and Chapman, C. J. 1982. Congenital defects and miscarriages among New Zealand 2,4,5-T sprayers. *Archives of Environmental Health*, Chicago, **37**: 197–200.

Thomas, H. F. 1980. 2,4,5-T use and congenital malformation rates in Hungary. *Lancet*, London, **1980**(2): 214–215.

Thomas, H. F. and Czeizel, A. 1982. Safe as 2,4,5-T? *Nature*, London, **295**: 276.

Townsend, J. C., Bodner, K. M., Van Peenen, P. F., Olsen, R. D. and Cook, R. R. 1982. Survey of reproductive events of wives of employees exposed to chlorinated dioxins. *American Journal of Epidemiology*, Baltimore, **115**: 695–713.

Westing, A. H. 1978. Ecological considerations regarding massive environmental contamination with 2,3,7,8-tetrachlorodibenzo-*para*-dioxin. *Ecological Bulletin*, Stockholm, **27**: 285–294.

Chapter Eight
Experimental Toxicology and Cytogenetics

8.A. Experimental toxicology and cytogenetics:
Symposium summary
Cung Bính Trung *et al.*

8.B. Chromosomal aberrations in humans following
exposure to herbicides
Cung Bính Trung and Vu Van Dieu

8.C. Experimental toxicology and cytogenetics:
an overview
Alastair W. M. Hay

8.A. Experimental toxicology and cytogenetics: Symposium summary[1]

Cung Bính Trung et al.[2]

College of Medicine, Hanoi

This summary report examines those aspects of experimental toxicology and cytogenetics relative to the long-term effects on human health of the herbicides applied in South Viet Nam during the Second Indochina War of 1961–1975. Of particular concern here is the dioxin contaminant of Agent Orange, the major herbicide that was employed.

Cytogenetics

Vietnamese scientists, using standard non-banding cytogenetic techniques and sister-chromatid exchange methods for investigations on chromosome aberrations, have reported an increase in chromosome aberrations and sister-chromatid exchanges in adults directly exposed to herbicides during the war in South Viet Nam as well as in their children (Trung & Dieu, paper 8.B). These people are still living in the sprayed area. A control group was selected from southern Viet Nam. The research was performed on the peripheral blood.

The abnormalities reported include chromatid breaks, chromosome breaks, translocations and polyploid cells. Some of these are rarely seen in human beings, especially the ring chromosomes, translocations with quadriradial figures and endo-reduplications. These aberrations have been found many years after the chemicals had been sprayed. Aberrations similar to those described have also been reported for victims of radiation exposure in Japan following the dropping of the two atomic bombs, and the Vietnamese scientists believe that their findings indicate that there has been a long-term health effect on the victims of the wartime herbicide exposure.

---

[1] Summary report of a Working Group of the International Symposium on Herbicides and Defoliants in War: the Long-term Effects on Man and Nature, Ho Chi Minh City, 13–20 January 1983 (appendix 3). During the course of the Symposium this Working Group and the one on 'Dioxin Chemistry' (paper 9.A) were combined in a formal sense, but functioned *de facto* largely as two groups.

[2] The Working Group consisted of: E. I. Astachkin, E. A. Carlson, A. W. M. Hay (Rapporteur), V. Hrdina, Z. Makles, S. Rump, Seng Lim Neou, Dang Nhu Tai, K. Tóth, R. Trapp, Cung Bính Trung (Chairman) and Bach Quoc Tuyên. See also the participants listed for the Working Group on 'Dioxin Chemistry' (paper 9.A, footnote 2).

The above information has been extended by other Vietnamese scientists who reported an increase in chromosome aberrations in spermatogonia and primary spermatocytes (both sperm cell precursors) caused by 2,4,5-T in *in vivo* tests on white mice.

Members of the Working Group discussed papers on dioxin that indicate a lack of mutagenicity (causing genetic damage) in fruit flies and, using the Ames test, in bacteria, but indicate the presence of mutations when the dioxin was tested in a mammalian cell transformation assay.

In the opinion of the Working Group, the cytogenetic investigations reported by the Vietnamese scientists are certainly interesting, but because of the controversial nature of the published literature on the subject further studies by additional laboratories are needed.

Toxicology

A member of the Working Group presented evidence on the carcinogenicity of dioxin in rodents. This study, considered alongside the five or so already published in the scientific literature, indicates that there is now sufficient evidence to class dioxin as a carcinogen in a number of animal species. It is not yet clear, on the other hand, whether dioxin acts directly or indirectly to cause cancer. However, the information presented here on the mutagenicity of dioxin in a cell transformation assay suggests that this chemical is an initiator and can cause cancer. Evidence was also presented for the carcinogenicity of the herbicide 2,4,5-trichlorophenoxyethanol in rodents.

A member of the Working Group presented evidence for the toxicity of herbicides in fruit flies. 2,4-D was toxic at a level of 1 000 mg/kg (ppm) whereas 2,4,5-T was toxic at a level of 300 ppm. The toxic effects included total failure of the life cycle of the fly at these doses; and a proportionate survival at lower doses, with a developmental delay which was not teratogenic, but which caused changes in the duration of the life cycle, the sex ratio of the emergent population, and the time of maturation. They also included behavioural modifications in the choice of media for egg laying. Media free of herbicides were preferred for egg laying over those containing 2,4-D, 2,4,5-T, or a mixture of the two. The dioxin content of the 2,4,5-T employed in these studies was not known.

The mode of action of the chlorinated pesticides, polychlorinated dibenzofurans and dioxins was discussed with reference to their action on the liver. Chemical warfare agents in general were reported to have delayed toxic effects in humans and it was suggested that a considerable research effort was required to find out more about this problem. In particular, it was suggested that workers employed in the manufacture of chemical weapons be studied for any long-term health problems.

Recommendations

The Working Group would like to see: (*a*) more *in vitro* studies using eukaryotic organisms (those whose cells possess nuclei) with different doses of herbicides in order to determine different frequencies of chromosome aberrations and gene mutations; (*b*) continued monitoring of the Vietnamese population exposed to the wartime herbicides in order to detect any mutagenic or carcinogenic effects in this and subsequent generations; and (*c*) co-operation among laboratories on an international basis in order to facilitate this work.

8.B. Chromosomal aberrations in humans following exposure to herbicides

Cung Bính Trung and Vu Van Dieu[1]

College of Medicine, Hanoi

I. Introduction

Large areas of forest and agricultural lands in South Viet Nam were attacked with herbicides during the Second Indochina War, especially from 1966 to 1969. As a result many people were directly or indirectly exposed to these agents of which Agent Orange (with its dioxin contaminant) made up more than 60 per cent in terms of volume sprayed.

The present study was aimed at assessing whether wartime exposure to the herbicides sprayed has had a lasting impact on the chromosomes of directly exposed individuals.

II. Materials and methods

The study was carried out during 1982 in two areas in southern Viet Nam: (*a*) a mountainous region in Dong Nai province (in the former Bien Hoa province); and (*b*) in a flat lowland region in Ben Tre province (in the former Kien Hoa province). A total of 66 persons (both males and females; average age, 38) was selected for examination, 36 from the mountainous region and 30 from the flat region. Among those from the mountainous region, 31 had been exposed to herbicides during the war and 5 had not; among those from the flat region, 25 had been exposed and 5 had not. The 56 exposed subjects had all been directly exposed to the spraying from one to five times during the period 1966–69, and all have continued to reside in a sprayed locale. Only individuals determined on the basis of clinical examination and interview to be free of diseases that result in chromosome aberrations and of not having used drugs with the same potential were selected as subjects.

The chromosomes investigated were those of the leucocytes (white blood cells) of the peripheral blood. The cells were cultured *in vitro* for 72 hours and then examined by standard procedures. An average of 106 cell nuclei at the metaphase

---

[1] Adapted by the editor from the authors' Symposium presentation.

stage of mitosis (division) were examined from each subject by light microscopy (in no case less than 80 per individual), and those exhibiting chromosomal aberrations photographed. Each such cell (i.e., photograph) was evaluated independently by three persons and their results averaged. Standard evaluation criteria were employed for the aberrations.

III. Results

It was found that the chromosome complement in the leucocytes of unexposed subjects differed from the normal value of 46 in 49 cells per thousand; by contrast, among those who had been exposed to herbicides during the war such numerical aberrations occurred in 111 cells per thousand, representing a greater than twofold increase (table 8.B.1). Among the polyploid cells of the exposed subjects about 10 per cent appeared to have attained this aberration through a failure of the nuclear membrane to disintegrate during mitosis, a phenomenon observed in only about 0.1 per cent of the polyploid cells of the control subjects.

In the leucocytes of unexposed subjects aberrant chromosomes were found in 19 cells per thousand, whereas in the previously herbicide-exposed subjects such structural aberrations were found in as many as 86 cells per thousand, that is, a greater than fourfold increase (table 8.B.1). Chromosome and chromatid

Table 8.B.1. Human chromosomal aberrations in southern Viet Nam, 1982

Chromosomal aberration	Without past spray exposure[a] (No./1 000)	With past spray exposure[b] (No./1 000)	Increase
Numerical aberrations	49.0	111	2.3 ×
Hypo-aneuploidy (<46)	38.4	75.8	2.0 ×
Hyper-aneuploidy (>46)	7.18	22.4	3.1 ×
Polyploidy (multiples of 46)	3.43	13.2	3.8 ×
Structural aberrations	18.7	85.7	4.6 ×
Gaps	7.80	14.9	1.9 ×
Breakage	0.625	10.0	16 ×
Chromatid breakage	3.43	40.2	12 ×
Two active centres (dicentric)	0	7.53	∞ ×
No active centre	6.87	8.09	1.2 ×
Translocation	0	3.48	∞ ×
Ring	0	1.46	∞ ×

[a] Based on 1 068 leucocytes from 10 subjects (average age, 37), 5 from Dong Nai province and 5 from Ben Tre province.
[b] Based on 5 934 leucocytes from 56 subjects (average age, 38), 31 from Dong Nai province and 25 from Ben Tre province.

(the product of a chromosome division) breakages accounted for most of this increase. The complement of 44 human chromosomes exclusive of the 2 sex chromosomes (i.e., the complement of autosomes) can be separated according to decreasing size into seven groups, A to G. It appears that among the exposed subjects a higher than expected proportion of the structural aberrations occur in groups B and C; and a lower than expected proportion in groups D, E and F.

IV. Conclusion

The present study suggests that previous exposure to herbicides can lead to an increased frequency of chromosomal aberrations, both of the numerical and gross structural types. The study was carried out more than a decade after the period of overt exposure, suggesting a long-lasting effect (or possibly long-lasting retention of the chemicals). It must be noted, however, that all of the test subjects have continued to reside in the previously sprayed locales, which suggests the possibility of continued exposure to any chemicals that might have remained in the environment.

An increase in chromosomal aberrations is a cause for concern in that such anomalies have in the literature been associated with increased frequencies of spontaneous abortion, congenital defects and leukaemia (the last especially in connection with hyper-aneuploidy).

8.C. Experimental toxicology and cytogenetics: an overview

Alastair W. M. Hay

University of Leeds, England

I. Introduction

There is growing concern about the long-term effects of exposure to herbicides. Nowhere is that concern more evident than in Viet Nam where the use of phenoxy herbicides was so intense and widespread during the Second Indochina War, especially from 1966 to 1969. If one such agent in particular should be singled out, it is Agent Orange, a 1:1 mixture of 2,4-D and 2,4,5-T. Agent Orange was also heavily contaminated with dioxin, a known teratogen (producer of congenital defects) and carcinogen (producer of cancer) in animals.

In view of the ability of dioxin to cause congenital defects and cancer in animals there is understandable concern that it may also act in a similar way in humans. The present overview examines the long-term health effects of phenoxy herbicides and their contaminants, in particular their potential mutagenic (causing genetic, i.e., DNA or gene, damage), chromosome-damaging and carcinogenic properties. The relevant published scientific literature is reviewed as well as some recent unpublished Vietnamese studies.

II. Mutagenicity

There is a widespread belief among cancer workers that DNA damage is involved in the induction of cancer (Bridges, 1976). In bladder cancer, for example, it appears that the mutation of a single gene (i.e., a single point mutation) is responsible for the development of the tumour (Reddy *et al.*, 1982; Tabin *et al.*, 1982).

The detection of chemical mutagens (agents that will damage DNA) is commonly accomplished by employing a short-term bacterial, fungal or mammalian cell bio-assay (Ames *et al.*, 1975; Bridges, 1976; Purchase *et al.*, 1978; 1982). Most mammalian mutagens require activation by certain liver enzymes, which are thus often added in these tests (Ames *et al.*, 1973). Most, but not all, chemical agents that test positive in these various bio-assays turn out to be carcinogenic in mammals; conversely, some that test negative turn out to be carcinogens as well

Dioxin has been tested using the above noted bio-assays, both with and without the presence of activating enzymes (Wassom et al., 1977–1978). Two groups of researchers have reported positive results without the addition of activator (Hussain et al., 1972; Seiler, 1973), whereas several groups have reported negative results either with or without activator (Geiger & Neal, 1981; Hay, 1982a, pages 41–47). Using qualitatively different mutagen bio-assays from the above ones, some groups have found dioxin to test at least weakly positive (Hay, 1982a, pages 41–47; Rogers et al., 1982).

Recently it has been suggested that the role of dioxin in carcinogenesis is only that of a promoter. Poland et al. (1982) have found that dioxin promotes the growth of skin tumours in mice that had been initiated by certain other mutagenic chemicals. This thesis is supported by the observation that dioxin does not bind appreciably to DNA in vivo (Poland & Glover, 1979).

Thus it appears clear from in vitro tests that dioxin is somehow involved in cancer induction. However, the exact mechanism of this carcinogenic action has not as yet been satisfactorily elucidated. The dioxin could be acting through its role as a mutagen, through its role as a promoter, or perhaps through both mechanisms. Moreover, the carcinogenic property suggested for dioxin by some of these mutagen bio-assays is well supported by direct evidence from animal studies and also by some indications from human studies (see below; see also Dwyer & Epstein, paper 6.D).

III. Chromosome damage

Under at least certain conditions, dioxin can be demonstrated to affect mitosis (cell division) and to cause chromosome aberrations. Thus in *Haemanthus katherinae* (African blood lily) dioxin is reported to have a powerful inhibitory effect on mitosis, to produce a number of chromosome aberrations (dicentric bridges, and so on) and to lead to multiple nuclei or a single large one (Jackson, 1972). Such multi-nucleated cells have also been observed in dioxin-treated mice (Vos et al., 1974) and other animals (Kimbrough et al., 1977).

Evidence of chromosome damage in humans exposed to herbicides has been put forth by a number of Vietnamese scientists.

In one study Trung & Dieu (paper 8.B) compared 56 adults residing in southern Viet Nam and directly exposed to herbicides one or more times during the Second Indochina War with a comparable unexposed control group. The authors report for the peripheral blood leucocytes of the exposed group increased incidences of both numerical aberrations (aneuploidy, polyploidy) and structural aberrations (chromosome breakages, chromatid [product of chromosome division] breakages, two active centres, and so on). For this investigation Trung & Dieu used a standard so-called non-banding technique. The now more routinely used banding techniques would have been able to disclose higher frequencies of structural aberrations in both the test and control groups, but it is unlikely that this would have altered their relative differences. It is a matter of some concern that an increased frequency of chromosome aberrations should be detected so many years (more than a decade) after direct exposure of a subject to herbicides. It will be necessary to find out whether the investigators were witnessing a long-lasting effect (a result not substantiated in other studies) or the result of chronic exposure to some chemical remaining in the environment of the previously sprayed region.

Bach Quoc Tuyên *et al.* (Bach Mai Hospital, Hanoi, unpublished) have studied the chromosomes in peripheral blood leucocytes of 27 individuals exposed to herbicides in South Viet Nam during the Second Indochina War, but who have been living in unsprayed northern Viet Nam since the war. This test group was compared with two control groups: (*a*) a presumably comparable group of 7 unexposed individuals who had also moved to the North; and (*b*) another presumably comparable group of 50 unexposed individuals who continued to live in the South. The exposed group was reported to have a higher frequency of chromosome aberrations than either of the two control groups. For example, chromatid breaks in the leucocytes of the exposed groups were said to be 23.2 per thousand as opposed to either 7.4 or 9.9, respectively, in the two control groups. Tuyên *et al.* reported as well that the incidence of chromosome aberrations was substantially higher in the offspring of women who had been exposed to the wartime herbicides during or even prior to their pregnancy. The work of Tuyên *et al.* clearly needs to be repeated in a carefully designed experiment, using a larger number of individuals, and with the subjects chosen by a method that avoids the possibility of bias (preferably by a so-called double-blind method).

Cung Bính Trung (College of Medicine, Hanoi, unpublished) recently examined the frequency of so-called sister-chromatid exchanges in peripheral blood leucocytes. This is a cryptic type of chromosome aberration considered to be a very sensitive indicator of exposure to chromosome-damaging chemicals. He compared the frequency of sister-chromatid exchanges in a group of 10 adults who had borne children with congenital defects and who had been exposed to the wartime herbicides with a group of 5 adults who had borne normal children and who had not been exposed. The frequency of this abnormality was reported to be about twice as high in the exposed group with defective children as in the unexposed group with normal children. The children of the exposed parents also seemed to have a higher than normal frequency of this abnormality. No conclusion can be drawn from this study owing to the small sample size and especially to its poor design and protocol.

IV. Carcinogenicity

It is not clear whether either 2,4-D or 2,4,5-T *per se* poses a carcinogenic risk to humans (IARC, 1982, pages 101, 211, 235). But the situation is quite different with dioxin: although the human data are still inconclusive, there is no doubt about its carcinogenicity in mammals (Hay, 1982b; IARC, 1982, page 238). At least four separate studies with rodents have shown that dioxin in the diet can cause cancer of the liver, lungs, thyroid, and so on (Van Miller et al., 1977; Kociba et al., 1978; National Toxicology Program, 1982a; Tóth et al., 1979). Another study with rodents has shown that application of dioxin to the skin can lead to cancer of the skin (National Toxicology Program, 1982b).

As suggested above, it is not yet clear whether dioxin is a carcinogen in humans. The interpretation of human studies is made difficult by the fact that everyone has been exposed to a variety of other chemicals as well. There have been a number of mortality studies of workers who had been occupationally exposed to phenoxy herbicides (with the level of dioxin contamination not known) in which no relationship could be discerned between such exposure and cancer incidence (Axelson & Sundell, 1974; Smith et al., 1982). However, in another such study (carried out in Sweden) Hardell & Sandström (1979) have reported a sixfold increase in the rare soft-tissue sarcomas.[1] Further work by Hardell and colleagues lends support to these findings (Hardell, 1981; Hardell et al., 1981; see also Dwyer & Epstein, paper 6.D; Vân, paper 6.C). Viet Nam— with its relatively few environmental carcinogens of industrial origin—would appear to provide an opportunity for investigating further the possible relationship in humans between exposure to phenoxy herbicides plus dioxin and the incidence of soft-tissue sarcomas.

V. Conclusion

Some in vitro tests suggest that dioxin, a contaminant associated with the phenoxy herbicide 2,4,5-T, is a mutagen and thus perhaps as well a mammalian (including human) carcinogen. Direct tests for dioxin carcinogenicity using rodents have confirmed that the chemical does cause cancer in animals, but whether it does so in humans has not as yet been established.

Disturbing indications exist from Viet Nam that exposure to phenoxy herbicides will cause chromosome aberrations. Supporting evidence for these reports is not available, however, from mammalian laboratory studies or from epidemiological surveys of occupationally exposed individuals in other countries. Careful follow-up studies are recommended.

---

[1] 'Soft-tissue sarcoma' is a fleshy malignant tumour of the soft somatic tissues (Morton & Eilber, 1982). The soft somatic tissues, which make up more than 50 per cent of the body weight, include the fibrous and adipose connective tissues, blood vessels, nerves, smooth and striated muscles, fascia (ligaments, tendons, and so on) and their synovial structures (sheaths and so on), and lymphatic structures. However, cancers of the lymphatic system—Hodgkin's disease (Rosenberg, 1982), Burkitt's lymphoma (Ziegler, 1982), and other malignant lymphomas (DeVita et al., 1982)—are not usually subsumed under the term 'soft-tissue sarcoma'.

References

Ames, B. N., Lee, F. D. and Durston, W. E. 1973. Improved bacterial test system for the detection and classification of mutagens and carcinogens. *Proceedings of the National Academy of Sciences*, Washington, **70**: 782–786.

Ames, B. N., McCann, J. and Yamasaki, E. 1975. Methods for detecting carcinogens and mutagens with the *Salmonella*/mammalian-microsome mutagenicity test. *Mutation Research*, Amsterdam, **31**: 347–363.

Axelson, O. and Sundell, L. 1974. Herbicide exposure, mortality and tumor incidence: an epidemiological investigation on Swedish railroad workers. *Work, Environment, Health*, Helsinki, **11**(1): 21–28.

Bridges, B. A. 1976. Short term screening tests for carcinogens. *Nature*, London, **261**: 195–200.

DeVita, V. T., Jr, Fisher, R. I., Johnson, R. E. and Berard, C. W. 1982. Non-Hodgkin's lymphomas. In: Holland, J. F. & Frei, E., III (eds). *Cancer Medicine*. Philadelphia: Lea & Febiger, 2465 pp.: pp. 1502–1537.

Geiger, L. E. and Neal, R. A. 1981. Mutagenicity testing of 2,3,7,8-tetrachlorodibenzo-*p*-dioxin in histidine auxotrophs of *Salmonella typhimurium*. *Toxicology and Applied Pharmacology*, New York, **59**: 125–129.

Hardell, L. 1981. Relation of soft-tissue sarcoma, malignant lymphoma and colon cancer to phenoxy acids, chlorophenols and other agents. *Scandinavian Journal of Work, Environment & Health*, Helsinki, **7**: 119–130.

Hardell, L. and Sandström, A. 1979. Case-control study: soft-tissue sarcomas and exposure to phenoxyacetic acids or chlorophenols. *British Journal of Cancer*, London, **39**: 711–717.

Hardell, L., Eriksson, M., Lenner, P. and Lundgren, E. 1981. Malignant lymphoma and exposure to chemicals, especially organic solvents, chlorophenols and phenoxy acids: a case-control study. *British Journal of Cancer*, London, **43**: 169–176.

Hay, A. 1982a. *Chemical Scythe: Lessons of 2,4,5-T and Dioxin*. London: Plenum Press, 275 pp.

Hay, A. 1982b. Phenoxy herbicides, trichlorophenols, and soft-tissue sarcomas. *Lancet*, London, **1982**(1): 1240.

Hussain, S., Ehrenberg, L., Löfroth, G. and Gejvall, T. 1972. Mutagenic effects of TCDD on bacterial systems. *Ambio*, Stockholm, **1**: 32–33.

IARC (International Agency for Research on Cancer). 1982. *IARC Monographs on the Evaluation of the Carcinogenic Risk of Chemicals to Humans. Suppl. No. 4*. Lyon: World Health Organization International Agency for Research on Cancer, 292 pp.

Jackson, W. T. 1972. Regulation of mitosis. III. Cytological effects of 2,4,5-trichlorophenoxy-acetic acid and of dioxin contaminants in 2,4,5-T formulations. *Journal of Cell Science*, New York, **10**: 15–25.

Kimbrough, R. D., Carter, C. D., Liddle, J. A., Cline, R. E. and Phillips, P. E. 1977. Epidemiology and pathology of a tetrachlorodibenzodioxin poisoning episode. *Archives of Environmental Health*, Chicago, **32**: 77–86.

Kociba, R. J. *et al.* 1978. Results of a two-year chronic toxicity and oncogenicity study of 2,3,7,8-tetrachlorodibenzo-*p*-dioxin in rats. *Toxicology and Applied Pharmacology*, New York, **46**: 279–303.

Morton, D. L. and Eilber, F. R. 1982. Soft tissue sarcomas. In: Holland, J. F. & Frei, E., III (eds). *Cancer Medicine*. Philadelphia: Lea & Febiger, 2465 pp.: pp. 2141–2157.

National Toxicology Program, US. 1982A. *Carcinogenesis Bioassay of 2,3,7,8-Tetrachloro-dibenzo-*p*-dioxin in Osborne–Mendel Rats and B6C3F$_1$ Mice: Gavage Study*. Research Triangle Park, N. Car.: US National Toxicology Program Technical Report Series No. 209, 195 pp. [Revision of Publ. No. (NIH)80-1765, 1980.]

National Toxicology Program, US. 1982b. *Carcinogenesis Bioassay of 2,3,7,8-tetrachloro-dibenzo-*p*-dioxin in Swiss–Webster Mice: Dermal Study*. Research Triangle Park, N. Car.: US National Toxicology Program Technical Report Series No. 201, 113 pp. [Revision of Publ. No. (NIH)80-1757, 1980.]

Poland, A. and Glover, E. 1979. Estimate of the maximum *in vivo* covalent binding of 2,3,7,8-tetrachlorodibenzo-*p*-dioxin to rat liver protein, ribosomal RNA, and DNA. *Cancer Research*, Philadelphia, **39**: 3341–3344.

Poland, A., Palen, D. and Glover, E. 1982. Tumour promotion by TCDD in skin of HRS/J hairless mice. *Nature*, London, **300**: 271–273.

Purchase, I. F. H. *et al.* 1978. Evaluation of 6 short-term tests for detecting organic chemical carcinogens. *British Journal of Cancer*, London, **37**: 873–959.

Purchase, I. F. H. *et al.* 1982. Evaluation of six short term tests for detecting organic chemical carcinogens and recommendations for their use. *Nature*, London, **264**: 624–627.

Reddy, E. P., Reynolds, R. K., Santos, E. and Barbacid, M. 1982. Point mutation is responsible for the acquisition of transforming properties by the T24 human bladder carcinoma oncogene. *Nature*, London, **300**: 149–152.

Rogers, A. M., Andersen, M. E. and Back, K. C. 1982. Mutagenicity of 2,3,7,8-tetrachloro-dibenzo-*p*-dioxin and perfluoro-*n*-decanoic acid in L5178Y mouse-lymphoma cells. *Mutation Research*, Amsterdam, **105**: 445–449.

Rosenberg, S. A. 1982. Hodgkin's disease. In: Holland, J. F. & Frei, E., III (eds). *Cancer Medicine*. Philadelphia: Lea & Febiger, 2465 pp.: pp. 1478–1502.

Seiler, J. P. 1973. Survey on the mutagenicity of various pesticides. *Experientia*, Basel, **29**: 622–623.

Smith, A. H., Fisher, D. O., Pearce, N. and Teague, C. A. 1982. Do agricultural chemicals cause soft tissue sarcoma?: Initial findings of a case-control study in New Zealand. *Community Health Studies*, Adelaide, **6**: 114–119.

Tabin, C. J. *et al.* 1982. Mechanism of activation of a human oncogene. *Nature*, London, **300**: 143–149.

Tóth, K., Somfai-Relle, S., Sugár, J. and Bence, J. 1979. Carcinogenicity testing of herbicide 2,4,5-trichlorophenoxyethanol containing dioxin and of pure dioxin in Swiss mice. *Nature*, London, **278**: 548–549.

Van Miller, J. P., Lalich, J. J. and Allen, J. R. 1977. Increased incidence of neoplasms in rats exposed to low levels of 2,3,7,8-tetrachlorodibenzo-*p*-dioxin. *Chemosphere*, Elmsford, N.Y., **6**(10): 537–544].

Vos, J. G., Moore, J. A. and Zinkl, J. G. 1974. Toxicity of 2,3,7,8-tetrachlorodibenzo-*p*-dioxin (TCDD) in C57B1/6 mice. *Toxicology and Applied Pharmacology*, New York, **29**: 229–241.

Wassom, J. S., Huff, J. E. and Loprieno, N. 1977–1978. Review of the genetic toxicology of chlorinated dibenzo-*p*-dioxins. *Mutation Research*, Amsterdam, **47**: 141–160.

Ziegler, J. L. 1982. Burkitt's lymphoma. In: Holland, J. F. & Frei, E., III (eds). *Cancer Medicine*. Philadelphia: Lea & Febiger, 2465 pp.: pp. 1537–1546.

Chapter Nine
Dioxin Chemistry

9.A. Dioxin chemistry : Symposium summary[1]

Tran Xuan Thu et al.[2]
University of Hanoi

This summary report examines those aspects of the chemistry of dioxin (TCDD; 2,3,7,8-tetrachlorodibenzo-*p*-dioxin) and closely related compounds relative to the long-term effects on human health of the herbicides applied in South Viet Nam during the Second Indochina War of 1961–1975. The dioxin was a contaminant of Agent Orange, the major herbicide that was employed.

According to official US figures, 44 million litres (57 million kilograms) of Agent Orange were expended, containing 22 million kilograms of 2,4-D (2,4-dichlorophenoxyacetic acid) and 24 million kilograms of 2,4,5-T (2,4,5-trichlorophenoxyacetic acid); 20 million litres (23 million kilograms), of Agent White were expended containing 5 million kilograms of 2,4-D and 1 million kilograms of picloram (4-amino-3,5,6-trichloropicolinic acid); and 8 million litres (11 million kilograms) of Agent Blue were expended, containing 3 million kilograms of cacodylic acid (dimethylarsinic acid; of which almost 2 million kilograms was elemental arsenic). In addition to these herbicides, the USA also expended some 9 million kilograms of the harassing agent CS (ortho-chlorobenzalmalononitrile). These amounts cannot be independently verified and some believe that they may have been higher.

The herbicides were sprayed onto at least 1.7 million hectares (and the harassing agent over perhaps an additional 5 million hectares). The concentration of herbicides used (in terms of the active moiety) varied from 15–20 kg/ha up to 300 kg/ha in unusual circumstances (with the average having been about 30 kg/ha).

Derived from information released by the USA, Agent Orange and its analogues contained an overall total of about 170 kg of dioxin. Some participants in the Working Group agreed that this amount of dioxin was approxi-

---

[1] Summary report of a Working Group of the International Symposium on Herbicides and Defoliants in War: the Long-term Effects on Man and Nature, Ho Chi Minh City, 13–20 January 1983 (appendix 3). During the course of the Symposium this Working Group and the one on 'Experimental Toxicology and Cytogenetics' (paper 8.A) were combined in a formal sense, but functioned *de facto* largely as two groups.

[2] The Working Group consisted of: G. Borissov, A. Chesnokov, Vu Ta Cuc, R. Dawa, A. V. Fokin, A. van der Gen, M. F. Kisseljov, A. F. Kolomietz, O. M. Lissov, K. Olie, T. N. Rao, C. Rappe (Rapporteur), Ho Si Thoang, Lê Van Thói and Tran Xuan Thu (Chairman). See also the participants listed for the Working Group on 'Experimental Toxicology and Cytogenetics' (paper 8.A, footnote 2).

mately correct. Based on analytical data from samples left over from the spraying programme in South Viet Nam and the amounts of 2,4,5-T produced in different factories in different years and subsequently sprayed, they also arrived at a total figure of about 170 kg. However, a majority of the Working Group came to the conclusion, based on some published information, that the total amount of dioxin was greater than 500 kg.

The Working Group suggested that owing to the high toxicity of the 2,3,7,8-TCDD isomer of the dioxin group and the large variation in toxicity among the different dioxin isomers, the analytical method used in dioxin analyses should have: (a) good reproducibility; (b) very high sensitivity (in the pg range); and (c) the ability to allow the quantification of specific isomers, especially of the 2,3,7,8-TCDD isomer.

To date, 2,3,7,8-TCDD has been found in several different types of samples, such as soil, sediment, vegetation, fish tissue, mammalian tissue and bovine milk; and also in human milk, blood, liver, kidney and adipose (fat) samples.

Although 2,3,7,8-TCDD is the major impurity found in Agent Orange, it should be pointed out that other dioxins such as 1,3,7,8-TCDD, 1,3,6,8-TCDD, 1,3,7-tri-CDD, 2,7-di-CDD and 2,8-di-CDD have also been reported, together with a series of dibenzofurans.

Dioxins have also been found in other technical products. Of special interest is the existence of 1 000 mg/kg (ppm) of TCDD isomers in diphenylether herbicides used in rice fields. The major isomers here are 1,3,6,8-TCDD and 1,3,7,9-TCDD, but other isomers have also been identified. However, the 2,3,7,8-TCDD isomer has not been found here.

The Working Group discussed the matter of secondary formation of dioxin after phenoxy herbicide spraying via photochemical or pyrolytic reactions. The environmental situation is very complex. However, experimental data do not indicate any extensive secondary formation of dioxin as it relates to the Viet Nam situation. Although the burning of 2,4,5-T salts results in the formation of some TCDD, the 2,4,5-T in Agent Orange was a butyl ester formulation.

A series of experiments has demonstrated the bio-availability of dioxin in soil and sediments. It is therefore recommended that tissue samples from both terrestrial and aquatic animals be analysed for dioxin.

The degradation of dioxin in soil is very slow. A half-life in soil of greater than 10 years has been reported.

The metabolism and/or excretion of dioxin in primates seems to be quite slow. The half-life in primates appears to be about one year. However, in small rodents the degradation is reported to be much faster.

Analysis for the dioxin content of the parent phenoxy herbicide can be carried out by standard methods of high-resolution gas chromatography (HRGC). However, the presence of trace levels should be confirmed by an additional technique such as mass spectroscopy.

Regarding Agent Blue, the Working Group noted that for the analysis of arsenic, atomic absorption and X-ray spectroscopy are the methods with the best sensitivity and reproducibility.

Recommendation

The Working Group is aware of only two analytical dioxin studies of samples from Viet Nam. In the first, fish and crustacean samples collected in South Viet Nam in 1970 were found by Baughman & Meselson (1973) to contain up to about 300 ng/kg (ppt) (dry weight) of dioxin. In the second, a recent study by Olie (paper 9.B), small amounts of dioxin (up to about 30 ppt) were identified in soil and sediment samples collected not long ago in southern Viet Nam. The Working Group recommends further analyses of critical samples of soil, sediments, fish and other aquatic animals, human milk, and human tissue samples. The first phase of such a project should include a brief screening of 'grab samples', followed by systematic sampling under the aegis of an international organization such as the United Nations Educational, Scientific and Cultural Organization (Unesco), United Nations Environment Programme (UNEP), or World Health Organization (WHO). After coding, the test samples, together with control samples, should be sent in a 'round robin' study to a number of different analytical laboratories in, for example, Amersterdam, Hanoi, Ho Chi Minh City, Lincoln (Nebraska, USA) and Umeå (Sweden).

Reference

Baughman, R. and Meselson, M. 1973. Analytical method for detecting TCDD (dioxin): levels of TCDD in samples from Vietnam. *Environmental Health Perspectives*, Research Triangle Park, N. Car., **1973**(5): 27–35.

9.B. Analysis for dioxin in soils of southern Viet Nam

Kees Olie[1]

University of Amsterdam

I. Introduction

Some 44 million litres of Agent Orange were sprayed onto an estimated 1.0 million hectares of forest and farm in South Viet Nam during the Second Indochina War, especially from 1966 to 1969 (Westing, 1976, chapter 3). The Agent Orange spraying represented about 60 per cent of the total volume and area that was sprayed during the war. This Agent Orange contained about 24 million kilograms of 2,4,5-T which, in turn, was contaminated with greater or lesser amounts of 2,3,7,8-tetrachlorodibenzo-*p*-dioxin, hereinafter referred to simply as 'dioxin', for an overall total of perhaps 170 kg (Westing, 1982, page 368; Young, 1983).

The extent to which the dioxin persists in the environment of southern Viet Nam remains a major public health concern because of the continuing potential for humans to be exposed to this extraordinarily toxic substance (see, e.g., Baughman & Meselson, 1973). In the present pilot study a number of recently collected soil samples from areas sprayed during the war were tested for the presence of dioxin in order to determine whether any of the dioxin has, indeed, persisted in the environment more than a decade after its application.

II. Materials and methods

Altogether 14 soil samples were collected from 11 locations in previously sprayed habitats of southern Viet Nam during the period 1980–1981, some from inland sites and others from the coastal mangrove forest habitat (table 9.B.1). There is a 60 per cent chance that the spraying had been with Agent Orange and not with one of the other, non-dioxin-containing, agents. Depth of sample was generally not recorded. The samples were air-dried and then kept at ambient temperatures in polyethylene bags or bottles until they were tested. No comparable control sample was collected from unsprayed areas.

Testing was carried out in 1981 and 1982. An amount of about 50 g from each of the 14 samples was dried at about 100°C and then extracted with benzene

---

[1] Adapted by the editor from the author's Symposium presentation.

Table 9.B.1. Recent dioxin levels in soils of southern Viet Nam sprayed during the Second Indochina War

Soil sample[a]	Location[b]	Depth (cm)	Dioxin content[c] (ng/kg)
1a	Inland	0?	16.4
b		50	16.2
2a	Coastal	0–10	<1
b		100	1.3
c		150	1.7
3	?	?	<1
4	?	?	<1
5	?	?	<1
6	?	?	<1
7	?	?	<1
8	?	?	<1
9	?	?	31.0
10	?	?	<1
11	?	?	20.6

[a] Soil samples supplied by Ton That Tung (Viet Duc Huu Nghi Hospital, Hanoi) in 1981 (soils 1 and 2 apparently collected in 1980; soils 3 to 11 apparently in 1981).
[b] Soil samples all collected from sites in southern Viet Nam that had been sprayed with some herbicidal agent during the Second Indochina War, that is, between 1962 and 1970. The coastal soil (soil 2) was collected in the Rung Sat region, that is, the mangrove habitat of the Saigon River delta, about 60 km south-east of Ho Chi Minh City.
[c] Dioxin (2,3,7,8-tetrachlorodibenzo-p-dioxin) content determined by the highly isomer-specific method of Lamparski & Nestrick (1982), which has a detection limit of about 1 ng/kg.

for 24 h. Following this each extract was concentrated to about 20 ml and subjected to a clean-up procedure for the purpose of removing substances that might mimic dioxin in the test procedure subsequently employed. For example, unstable compounds were removed in one fashion, humic acids and sulphur in another, and polychlorinated biphenyls and chlorinated benzenes in still another. The sample was then evaporated to dryness and redissolved in 5 μL of nonane. Determination of the dioxin content of this moiety was carried out using a gas chromatograph coupled to a mass spectrometer, according to the '2,3,7,8' isomer-specific method of Lamparski & Nestrick (1982). The lowest amount of dioxin that can be detected by this method in a sample of soil is about 1 ng/kg.

III. Results

Of the 11 soils examined 4 tested positive for dioxin, the concentrations ranging from a low of about 1 ng/kg (the limit of detection) to a high of 31 ng/kg (table 9.B.1).

174

IV. Conclusion

The results indicate that the dioxin introduced into the environment of South Viet Nam during the Second Indochina War has, indeed, persisted to this day. Moreover, it is possible to calculate from the present findings a very crude approximation of the environmental half-life of dioxin in Viet Nam. For the initial point one can make the much simplified assumption that the soil concentration was 64 ng/kg in 1968. This amount represents half the average total amount applied during the war, distributed within the top 10 cm of the soil (and assuming an average soil weight of 1 325 kg/m³). Dioxin generally exhibits very little downward mobility in soil; greater or lesser amounts of lateral movement do occur, being the result of erosional movement (via water and wind) of soil particles to which the dioxin is bound (adsorbed). Only half the total is being used here in order to account for partial photo-degradation shortly after spraying and prior to soil incorporation (Crosby & Wong, 1977). The date represents the high point of the wartime spraying. For the final point one can use 10 ng/kg in 1981. This amount represents the average soil concentration of the soils tested, taking into account that probably only 60 per cent of the sites sampled had been sprayed with dioxin-containing Agent Orange; the date represents the primary year of sampling. Using these two sets of values and assuming exponential decay, the environmental half-life comes to 5 years. This estimate compares with two based on a field study in Florida, one of 3.5 years (Westing, 1982, page 368) and another of 10–12 years (Young, 1983).

Finally it must be noted that the data presented in this paper on persistence of dioxin in the soils of southern Viet Nam do not permit predictions about dioxin concentrations in the indigenous plants, animals or humans.

References

Baughman, R. and Meselson, M. 1973. Analytical method for detecting TCDD (dioxin): levels of TCDD in samples from Vietnam. *Environmental Health Perspectives*, Research Triangle Park, N. Car., **1973**(5): 27–35.

Crosby, D. G. and Wong, A. S. 1977. Environmental degradation of 2,3,7,8-tetrachlorodibenzo-*p*-dioxin. *Science*, Washington, **195**: 1337–1338.

Lamparski, L. L. and Nestrick, T. J. 1982. Isomer-specific determination of tetrachlorodibenzo-*p*-dioxin at part per trillion concentrations. In: Hutzinger, O., Frei, R. W., Merian, E. & Pocchiari, F. (eds). *Chlorinated Dioxins and Related Compounds: Impact on the Environment.* Oxford: Pergamon Press, 658 pp.: pp. 1–13.

Westing, A. H. 1976. In: SIPRI, *Ecological Consequences of the Second Indochina War.* Stockholm: Almqvist & Wiksell, 119 pp. + 8 pl.

Westing, A. H. 1982. Environmental aftermath of warfare in Viet Nam. In: *World Armaments and Disarmament, SIPRI Yearbook 1982.* London: Taylor & Francis, 518 pp.: pp. 363–389.

Young, A. L. 1983. Long-term studies on the persistence and movement of TCDD in a natural ecosystem. In: Tucker, R. E., Young, A. L. & Gray, A. P. (eds). *Human and Environmental Risks of Chlorinated Dioxins and Related Compounds.* New York: Plenum Press, 823 pp.: pp. 173–190.

9.C. Dioxin chemistry: an overview

Christoffer Rappe

University of Umeå, Sweden

I. Introduction

The polychlorinated dibenzo-*p*-dioxins represent a series of tricylcic aromatic compounds which exhibit similar physical and chemical properties and, owing to their toxic nature, have been the subject of much public health concern. The dioxins have been involved in a number of celebrated releases to the environment, for example, in Missouri, USA in 1971 (Carter *et al.*, 1975; Kimbrough *et al.*, 1977), in Sèveso, Italy in 1976 (Homberger *et al.*, 1979), and—the focus of the present overview—in South Viet Nam primarily from 1966 to 1969 (Westing, 1976, chapter 3).

The number of chlorine atoms in the chlorinated dibenzo-*p*-dioxins can vary between one and eight, and therefore altogether 75 such isomers can exist (figure 9.C.1). There is a pronounced difference in toxic effects among the different isomers, the most acutely toxic one being 2,3,7,8-tetrachlorodibenzo-*p*-dioxin. The '2,3,7,8' isomer is lethal to a number of sensitive laboratory mammals at dosages of less than 100 μg/kg, and to guinea-pigs (the most sensitive of the various mammals so far tested) of less than 1 μg/kg. The '2,3,7,8' isomer also has carcinogenic (cancer-causing) and teratogenic (birth defect-causing) properties in mammals and perhaps humans (Westing, paper 7.C; Dwyer & Epstein,

Figure 9.C.1. Dioxin structure[a]

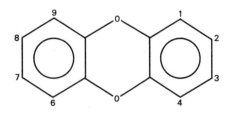

[a] The structure depicted is that of dibenzo-*p*-dioxin. One or more of the 8 numbered positions can be occupied by a chlorine atom (Cl). Thus, 75 chlorinated isomers exist: 2 with 1 Cl; 10 with 2 Cl; 14 with 3 Cl; 22 with 4 Cl; 14 with 5 Cl; 10 with 6 Cl; 2 with 7 Cl; and 1 with 8 Cl. The '2,3,7,8' isomer (i.e., 2,3,7,8-tetrachlorodibenzo-*p*-dioxin) is the most highly toxic one by far.

paper 6.D; Hay, paper 8.C). Other toxic dioxins include especially the '1,2,3,7,8', '1,2,3,4,7,8', '1,2,3,6,7,8' and '1,2,3,7,8,9' isomers.

II. Analytical procedures

Very sensitive and highly specific analytical techniques are required owing to the extreme toxicity of some of the dioxins, especially the '2,3,7,8' isomer. Detection levels in biological and environmental samples must be orders of magnitude below the usual ones required for the herbicides themselves. Indeed, a sensitivity of 1 ng/kg is necessary for the crucial '2,3,7,8' isomer. Attaining satisfactory specificity (avoidance of either false positive or false negative results) becomes very difficult at such low levels. Prerequisites for good analysis are: proper sampling; good storage; efficient extraction; satisfactory purification (clean-up and containment); excellent isomer separation; sensitive detection; and adequate confirmation. It is of major importance to be able to distinguish among the various isomers owing to the wide range of their toxic effects, and such techniques are only now becoming available (Buser & Rappe, 1980; n.d.).

The most sensitive and most specific techniques for dioxin analysis are based on gas chromatography separation and mass spectrometry isomer identification (Buser & Rappe, 1980; n.d.; Karasek & Onuska, 1982; McKinney, 1978; Rappe & Buser, 1980). The isomers can be analysed all in one fraction (unseparated) by a so-called containment procedure (Norstrom et al., 1982; Stalling et al., 1983). Conversely, a few of the isomers—especially the '2,3,7,8' isomer—can be analysed following their fractionation (separation) by a so-called high-performance liquid chromatography procedure (Lamparski et al., 1979). Quantification is carried out by a procedure referred to as mass-specific detection or mass fragmentography (Rappe et al., 1983a).

III. Levels in environmental samples

Direct introduction

Dioxins can enter the environment from a number of sources: as an impurity in a number of industrial chemicals that are either intentionally applied or discarded outdoors; as a component of smoke emissions from municipal incinerators and heating plants; as a component of chemicals that are unintentionally dispersed through factory explosions or other accidents; and in other ways. The dioxins that entered the environment of South Viet Nam during the Second Indochina War were an impurity in the 2,4,5-T component of Agent Orange which had been widely applied to agricultural and forest lands for hostile herbicidal purposes. The amount of the '2,3,7,8' isomer in Agent Orange varied from lot to lot and most of these are no longer available for analysis.

On the basis of an exceedingly small sample of salvaged late-war lots—the only ones still available—the total amount introduced into South Viet Nam has been estimated to be about 170 kg (Westing, 1982, page 368; Young, 1983; Young et al., 1978; see also Rappe et al., 1978).

It has been stated that the half-life in soil of the '2,3,7,8' isomer of dioxin might be as lengthy as 10–12 years (Young, 1983), although the data on which this estimate is based suggest a value of about 3–4 years (Westing, 1982, page 368). Uptake from the soil by vegetation, however, seems to be very slight. Vegetation analysed immediately following the Seveso accident was found to have rather high levels of this isomer of dioxin (up to 50 mg/kg), presumably via direct above-ground contact (Firestone, 1978). In the following years the levels of dioxin in the vegetation that grew subsequent to the accident dropped by several orders of magnitude and, moreover, were undetectable (down to a detection limit of 1 ng/kg) in the flesh of apples, pears, peaches, or corn (maize) (Wipf & Schmid, 1983). Fruit skins did contain about 100 ng/kg, suggesting contamination via dust (the soil at that time containing 10 μg/kg).

Sundström et al. (1979) sprayed vegetation with 2,4,5-T containing 60 μg/kg of the '2,3,7,8' isomer of dioxin. This amounted to a dioxin application to the leaves of about 600 ng/kg. A sample of leaves collected 42–45 days after the spraying was found to have 170 ng/kg of dioxin. The about 70 per cent loss of dioxin was presumably largely the result of photochemical degradation.

Secondary formation

The question arises from time to time of the extent to which dioxins—especially the '2,3,7,8' isomer—can be generated in the environment, for example, via photochemical or thermal reactions involving 2,4,5-T or other phenoxy herbicides.

The photochemical formation of the '2,3,7,8' isomer of dioxin from phenoxy herbicides has been studied under laboratory conditions by Åkermark (1978). Although it was possible to detect very small amounts of this dioxin, one could expect that under environmental conditions the dioxin would be efficiently photo-dechlorinated to much less toxic products owing to the presence of the phenoxy herbicides (inasmuch as these are potent hydrogen donors). Another possibility for the photochemical formation of the '2,3,7,8' isomer is via the dechlorination of more highly chlorinated isomers. Thus, Crosby et al. (1971) reported that 'octachloro' dioxin could be photo-degraded to 'tetrachloro' dioxin. However, this reaction has subsequently been shown not to produce the '2,3,7,8' isomer, but rather the substantially less toxic '1,4,6,9' isomer (Buser & Rappe, 1978).

The formation of the '2,3,7,8' isomer of dioxin as a result of thermal reactions of 2,4,5-T or related compounds has been suggested by the report of Buu-Hoï et al. (1971). However, Langer et al. (1973) pointed out errors in this study and themselves found that when salts of 2,4,5-T (but not the pure acid or its esters) were heated to 400–450°C for 30 min, then the '2,3,7,8' isomer was generated to the extent of 1 g/kg. Moreover, using a more sensitive analytical method, Ahling et al. (1977) reported generation of the '2,3,7,8' isomer from esters of 2,4,5-T during combustion at 500–580°C, to the extent of 0.2–3.0 mg/kg. Perhaps more to the point at hand, in a study where vegetation was burned at 600°C after having been experimentally treated with 2,4,5-T, no '2,3,7,8' isomer could be detected in the combustion gases, soot, or ashes (down to a detection limit of 4 mg/kg) (Rappe, 1978, pages 23–24). However, in a similar study, Stehl & Lamparski (1977) reported a conversion of 2,4,5-T to the '2,3,7,8' isomer of 1.6 mg/kg when grass which had been sprayed with 2,4,5-T was burned.

IV. Conclusion

A number of fish (4) and shellfish (1) collected in South Viet Nam in 1970 from the vicinity of previously heavily sprayed areas were reported to contain 2,3,7,8-tetrachlorodibenzo-p-dioxin in the range of 10 to 300 ng/kg (dry weight) (Baughman & Meselson, 1973), although it should be noted that the analytical technique that was employed is not isomer specific. Recent preliminary analyses of soils from southern Viet Nam that had been sprayed during the war have revealed the continued presence of the '2,3,7,8' isomer, at levels of up to about 30 ng/kg (Olie, paper 9.B). A recent study of individuals occupationally exposed to dioxin has suggested that a good correlation exists between duration of exposure and level in the blood plasma (Rappe et al., 1983b), and Shepard & Young (1983) have alluded to the presence of low levels of 'tetrachloro' dioxin (believed to be the '2,3,7,8' isomer) in adipose (fat) tissue of US veterans of the Second Indochina War.

Considering the above information, it is highly recommended that further research on critical samples from Viet Nam be performed. These should include samples of soil, sediment, fin-fish and shellfish, human milk, and human tissue. Such sampling and testing ought to be carried out systematically under the auspices of some international organization—for example, the United Nations Educational, Scientific and Cultural Organization (Unesco), United Nations Environment Programme (UNEP), or World Health Organization (WHO)—using carefully defined protocols and with replications in different laboratories.

References

Ahling, B., Lindskog, A., Jansson, B. and Sundström, G. 1977. Formation of polychlorinated dibenzo-p-dioxins and dibenzofurans during combustion of a 2,4,5-T formulation. *Chemosphere*, Elmsford, N.Y., **6**(8): 461–468.
Åkermark, B. 1978. Photochemical reactions of phenoxy acids and dioxins. *Ecological Bulletin*, Stockholm, **1978**(27): 75–81.
Baughman, R. and Meselson, M. 1973. Analytical method for detecting TCDD (dioxin): levels of TCDD in samples from Vietnam. *Environmental Health Perspectives*, Research Triangle Park, N. Car., **1973**(5): 27–35.

Buser, H. R. and Rappe, C. 1978. Identification of substitution patterns in polychlorinated dibenzo-*p*-dioxins (PCDDs) by mass spectrometry. *Chemosphere*, Elmsford, N.Y., **7**(2): 199–211.

Buser, H. R. and Rappe, C. 1980. High-resolution gas chromatography of the 22 tetra-chlorodibenzo-*p*-dioxin isomers. *Analyical Chemistry*, Washington, **52**: 2257–2262.

Buser, H. R. and Rappe, C. n.d. Isomer-specific separation of 2,3,7,8-substituted poly-chlorinated dibenzo-*p*-dioxins (PCDDs) using high-resolution gas chromatography and mass spectrometry. *Analytical Chemistry*, Washington, in the press.

Buu-Hoï, N. P., Saint-Ruf, G., Bigot, P., and Mangane, M. 1971. [Preparation, properties, and identification of the "dioxin" (2,3,7,8-tetrachlorodibenzo-*p*-dioxin) in the pyrolyzates of defoliants based on 2,4,5-trichlorophenoxyacetic acid and its esters and of contaminated vegetation.] (In French) *Comptes Rendus des Séances de l'Académie des Sciences (Série D)*, Paris, **273**: 708–711.

Carter, C. D. *et al.* 1975. Tetrachlorodibenzodioxin: an accidental poisoning episode in horse arenas. *Science*, Washington, **188**: 738–740.

Crosby, D. G., Wong, A. S., Plimmer, J. R. and Woolson, F. A. 1971. Photodecomposition of chlorinated dibenzo-*p*-dioxins. *Science*, Washington, **173**: 748–749.

Firestone, D. 1978. 2,3,7,8-tetrachlorodibenzo-*para*-dioxin problem: a review. *Ecological Bulletin*, Stockholm, **1978**(27): 39–52.

Homberger, E., Reggiani, G., Sambeth, J. and Wipf, H. K. 1979. Seveso accident: its nature, extent and consequences. *Annals of Occupational Hygiene*, Oxford, **22**: 327–370.

Karasek, F. W. and Onuska, F. I. 1982. Trace analysis of the dioxins. *Analytical Chemistry*, Washington, **54**: 309A–324A.

Kimbrough, R. D., Carter, C. D., Liddle, J. A., Cline, R. E. and Phillips, P. E. 1977. Epi-demiology and pathology of a tetrachlorodibenzodioxin poisoning episode. *Archives of Environmental Health*, Chicago, **32**: 77–86.

Lamparski, L. L., Nestrick, T. J. and Stehl, R. H. 1979. Determination of part-per-trillion concentrations of 2,3,7,8-tetrachlorodibenzo-*p*-dioxin in fish. *Analytical Chemistry*, Washington, **51**: 1453–1458.

Langer, H. G., Brady, T. P. and Briggs, P. R. 1973. Formation of dibenzodioxins and other condensation products from chlorinated phenols and derivatives. *Environmental Health Perspectives*, Research Triangle Park, N. Car., **1973**(5): 3–7.

McKinney, J. D. 1978. Analysis of 2,3,7,8-tetrachlorodibenzo-*para*-dioxin in environmental samples. *Ecological Bulletin*, Stockholm, **1978**(27): 53–66.

Norstrom, R. J., Hallett, D. J., Simon, M. and Mulvihill, M. J. 1982. Analysis of Great Lakes herring gull eggs for tetrachlorodibenzo-*p*-dioxins. In: Hutzinger, O., Frei, R. W., Merian, E. & Pocchiari, F. (eds.). *Chlorinated Dioxins and Related Compounds: Impact on the Environment*. Oxford: Pergamon Press, 658 pp.: pp. 173–181.

Rappe, C. 1978. Chlorinated phenoxy acids and their dioxins: chemistry: summary. *Ecological Bulletin*, Stockholm, **1978**(27): 19–27.

Rappe, C. and Buser, H. R. 1980. Chemical properties and analytical methods. In: Kimbrough, R. (ed.). *Halogenated Biphenyls, Terphenyls, Naphthalenes, Dibenzodioxins and Related Products*. Amsterdam: Elsevier/North-Holland Biomedical Press, 406 pp.: pp. 41–76.

Rappe, C., Buser, H. R. and Bosshardt, H.-P. 1978. Identification and quantification of poly-chlorinated dibenzo-*p*-dioxins (PCDDs) and dibenzofurans (PCDFs) in 2,4,5-T-ester formu-lations and Herbicide Orange. *Chemosphere*, Elmsford, N.Y., **7**(5): 431–438.

Rappe, C., Marklund, S., Nygren, M. and Garå, A. 1983a. Parameters for identification and confirmation in trace analyses of polychlorinated dibenzo-*p*-dioxins and dibenzofurans. In: Choudhary, G., Keith, L. H. & Rappe, C. (eds). *Chlorinated Dioxins and Dibenzofurans in the Total Environment*. Boston, Mass.: Butterworths, 416 pp.: pp. 259–272.

Rappe, C., Nygren, M. and Gustafsson, G. 1983b. Human exposure to polychlorinated dibenzo-*p*-dioxins and dibenzofurans. In: Choudhary, G., Keith, L. H. & Rappe, C. (eds). *Chlorinated Dioxins and Dibenzofurans in the Total Environment*. Boston, Mass.: Butter-worths, 416 pp.: pp. 355–365.

Shepard, B. M. and Young, A. L. 1983. Dioxins as contaminants of herbicides: the U.S. perspective. In: Tucker, R E., Young, A. L. & Gray, A. P. (eds). *Human and Environmental Risks of Chlorinated Dioxins and Related Compounds*. New York: Plenum Press, 823 pp.: pp. 3–11.

Stalling, D. L. *et al.* 1983. Residues of polychlorinated dibenzo-*p*-dioxins and dibenzofurans in Laurentian Great Lakes fish. In: Tucker, R. E., Young, A. L. & Gray, A. P. (eds). *Human and Environmental Risks of Chlorinated Dioxins and Related Compounds.* New York: Plenum Press, 823 pp.: pp. 221–240.

Stohl, R. H. and Lamparski, L. L. 1977. Combustion of several 2,4,5-trichlorophenoxy compounds: formation of 2,3,7,8-tetrachlorodibenzo-*p*-dioxin. *Science*, Washington, **197**: 1008–1009.

Sundström, G., Jensen, S., Jansson, B. and Erne, K. 1979. Chlorinated phenoxyacetic acid derivatives and tetrachlorodibenzo-*p*-dioxin in foliage after application of 2,4,5-trichlorophenoxyacetic acid esters. *Archives of Environmental Contamination & Toxicology*, New York, **8**: 441–448.

Westing, A. H. 1976. In: SIPRI, *Ecological Consequences of the Second Indochina War.* Stockholm: Almqvist & Wiksell, 119 pp. + 8 pl.

Westing, A. H. 1982. Environmental aftermath of warfare in Viet Nam. In: *World Armaments and Disarmament, SIPRI Yearbook 1982.* London: Taylor & Francis, 518 pp.: pp. 363–389.

Wipf, H. K. and Schmid, J. 1983. Seveso: an environmental assessment. In: Tucker, R. E., Young, A. L. & Gray, A. P. (eds). *Human and Environmental Risks of Chlorinated Dioxins and Related Compounds.* New York: Plenum Press, 823 pp.: 255–274.

Young, A. L. 1983. Long-term studies on the persistence and movement of TCDD in a natural ecosystem. In: Tucker, R. E., Young, A. L. & Gray, A. P. (eds). *Human and Environmental Risks of Chlorinated Dioxins and Related Compounds.* New York: Plenum Press, 823 pp.: pp. 173–190.

Young, A. L., Calcagni, J. A., Thalken, C. E. and Tremblay, J. W. 1978. *Toxicology, Environmental Fate, and Human Risk of Herbicide Orange and Its Associated Dioxin.* San Antonio, Tex.: US Air Force Occupational & Environmental Health Laboratory Rept No. OEHL TR-78-92, 247 pp.

Appendices

1. Bibliography
Arthur H. Westing

2. Scientific names of biota

3. International Symposium on Herbicides and Defoliants in War: the Long-term Effects on Man and Nature, Ho Chi Minh City, 13–20 January 1983

4. Overall Symposium summary
Hoàng Dình Câu *et al.*

Appendix 1. Bibliography

Arthur H. Westing

Stockholm International Peace Research Institute

Beljan, J. R. *et al.* 1981. *Health Effects of "Agent Orange" and Polychlorinated Dioxin Contaminants: Technical Report*. Chicago: American Medical Association, 39 pp.

Buckingham, W. A., Jr. 1982. *Operation Ranch Hand: The Air Force and Herbicides in Southeast Asia 1961–1971*. Washington: US Air Force Office of Air Force History, 253 pp.

Choudhary, G., Keith, L. H. and Rappe, C. (eds). 1983. *Chlorinated Dioxins and Dibenzofurans in the Total Environment*. Boston, Mass.: Butterworths, 416 pp.

Coulston, F. and Pocchiari, F. (eds). 1983. *Accidental Exposure to Dioxins: Human Health Aspects*. New York: Academic Press, 294 pp.

Diaz-Colon, J. D. and Bovey, R. W. 1978. *Selected Bibliography of Phenoxy Herbicides. VII. Military Uses*. College Station, Tex.: Texas Agricultural Experiment Station Publ. No. MP-1387, 26 pp.

Epstein, S. S. 1983. Agent Orange diseases: problems of causality, burdens of proof and restitution. *Trial*, Cambridge, Mass., **19**: 91–99, 137–138.

Hay, A. 1982. *Chemical Scythe: Lessons of 2,4,5-T and Dioxin*. London: Plenum Press, 275 pp.

Hay, J. R. *et al.* 1978. *Phenoxy Herbicides: Their Effects on Environmental Quality with Accompanying Scientific Criteria for 2,3,7,8-Tetrachlorodibenzo-p-dioxin (TCDD)*. Ottawa: Canada National Research Council Publ. No. NRCC 16075, 440 pp.

Homberger, E., Reggiani, G., Sambeth, J. and Wipf, H. K. 1979. Seveso accident: its nature, extent and consequences. *Annals of Occupational Hygiene*, Oxford, **22**: 327–370.

Hutzinger, O., Frei, R. W., Merian, E. and Pocchiari, F. (eds). 1982. *Chlorinated Dioxins and Related Compounds: Impact on the Environment*. Oxford: Pergamon Press, 658 pp.

Jessop, D. S. *et al.* 1982. *Pesticides and the Health of Australian Vietnam Veterans: First Report*. Canberra: Commonwealth of Australia Parliament Senate Standing Committee on Science and the Environment, 240 pp.

JRB Associates. 1981. *Review of Literature on Herbicides, including Phenoxy Herbicides and Associated Dioxins. I. Analysis of Literature. II. Annoted Bibliography*. Washington: US Veterans Administration Dept of Medicine & Surgery, 325 + 400 pp.

Kimbrough, R. (ed.). 1980. *Halogenated Biphenyls, Terphenyls, Naphthalenes, Dibenzodioxins and Related Products*. Amsterdam: Elsevier/North-Holland Biomedical Press, 406 pp.

Kimbrough, R. D., Carter, C. D., Liddle, J. A., Cline, R. E. and Phillips, P. E. 1977. Epidemiology and pathology of a tetrachlorodibenzodioxin poisoning episode. *Archives of Environmental Health*, Chicago, **32**: 77–86.

Kunstadter, P. 1982. *Study of Herbicides and Birth Defects in the Republic of Vietnam: An Analysis of Hospital Records*. Washington: National Academy Press, 73 pp.

Lang, A. *et al.* 1974. *Effects of Herbicides in South Vietnam. A. Summary and Conclusions*. Washington: National Academy of Sciences, [398] pp. + 8 maps.

Lathrop, G. D., Moynahan, P. M., Albanese, R. A. and Wolfe, W. H. 1983. *Project Ranch Hand II: An Epidemiologic Investigation of Health Effects in Air Force Personnel Following Exposure to Herbicides: Baseline Mortality Study Results*. San Antonio, Tex.: US Air Force, Brooks Air Force Base School of Aerospace Medicine, Epidemiology Division, 61 pp.

Lathrop, G. D., Wolfe, W. H., Albanese, R. A. and Moynahan, P. M. 1984. *Air Force Health Study (Project Ranch Hand II): An Epidemiologic Investigation of Health Effects in Air Force Personnel Following Exposure to Herbicides: Baseline Morbidity Study Results*. San Antonio, Tex.: US Air Force, Brooks Air Force Base School of Aerospace Medicine, Aerospace Medical Division. [359] pp.

McConnell, E. E. *et al.* 1984. Dioxin in soil: bioavailability after ingestion by rats and guinea pigs. *Science*, Washington, **223**: 1077–1079.

Moore, J. A. (ed.). 1973. Perspective on chlorinated dibenzodioxins and dibenzofurans. *Environmental Health Perspectives*, Research Triangle Park, N. Car., **1973**(5): 1–312.

Moore, J. A. *et al.* 1978. *Long-term Hazards of Polychlorinated Dibenzodioxins and Polychlorinated Dibenzofurans*. Lyon: World Health Organization International Agency for Research on Cancer Technical Rept No. 78/0001, 57 pp.

Norris, L. A. 1981. Movement, persistence, and fate of the phenoxy herbicides and TCDD in the forest. *Residue Reviews*, New York, **80**: 65–135.

Page, W. F., Gee, S. C. and Kuntz, A. J. 1983. *Protocol for the Vietnam Veteran Mortality Study*. Washington: US Veterans Administration Office of Reports and Statistics, 111 pp.

Poland, A. and Knutson, J. C. 1982. 2,3,7,8-tetrachlorodibenzo-*p*-dioxin and related halogenated aromatic hydrocarbons: examination of the mechanism of toxicity. *Annual Review of Pharmacology & Toxicology*, Palo Alto, Cal., **22**: 517–554.

Ramel, C. (ed.). 1978. Chlorinated phenoxy acids and their dioxins: mode of action, health risks and environmental effects. *Ecological Bulletin*, Stockholm, **1978**(27): 1–302.

Staats, E. B. 1979. *U.S. Ground Troops in South Vietnam Were in Areas Sprayed with Herbicide Orange*. Washington: US General Accounting Office Rept No. FPCD-80-23, 9 + 12 pp.

Tucker, R. E., Young, A. L. and Gray, A. P. (eds). 1983. *Human and Environmental Risks of Chlorinated Dioxins and Related Compounds*. New York: Plenum Press, 823 pp.

Westing, A. H. 1974. *Herbicides as Weapons: A Bibliography*. Santa Barbara, Cal.: California State Univ. Center for the Study of Armament and Disarmament Political Issue Series No. 3(1), 36 pp.

Westing, A. H. 1976. In: SIPRI, *Ecological Consequences of the Second Indochina War*. Stockholm: Almqvist & Wiksell, 119 pp. + 8 pl.

Westing, A. H. 1982. Environmental aftermath of warfare in Viet Nam. In: *World Armaments and Disarmament, SIPRI Yearbook 1982*. London: Taylor & Francis, 518 pp.: pp. 363–389.

Young, A. L., Calcagni, J. A., Thalken, C. E. and Tremblay, J. W. 1978. *Toxicology, Environmental Fate, and Human Risk of Herbicide Orange and Its Associated Dioxin*. San Antonio, Tex.: US Air Force Occupational & Environmental Health Laboratory Rept No. OEHL TR-78-92, 247 pp.

Young, A. L., Kang, H. K. and Shepard, B. M. 1983. Chlorinated dioxins as herbicide contaminants. *Environmental Science & Technology*, Washington, **17**: 530A–540A.

Appendix 2. Scientific names of biota

Plants

Acacia aneura
  Leguminosae
  Keo bong vàng
Acacia auriculaeformis
  Leguminosae
Acacia mangium
  Leguminosae
Acrostichum aureum
  Polypodiaceae
  Ráng dai
Adina sessilifolia (= *sessiliflora*)
  Rubiaceae
  Gáo vàng
Albizia falcataria
  Leguminosae
Albizia lebbek
  Leguminosae
  Hop hoan
Anisoptera
  Dipterocarpaceae
  Vên vên nam
Anisoptera costata (= *cochinchinensis*)
  Dipterocarpaceae
  Vên vên nam
Artocarpus
  Moraceae
  Jack fruit, mít
Avicennia
  Verbenaceae
  Mam
Avicinia marina
  Verbenaceae
  Mam den

Bruguiera
  Rhizophoraceae
  Vet
Bruguiera sexangula
  Rhizophoraceae
  Vet den

Cassava see *Manihot*
Cassia
  Leguminosae
  Muong

Casuarina
  Casuarinaceae
Ceriops
  Rhizophoraceae
Ceriops tagal
  Rhizophoraceae
  Nét
Colona auriculata
  Tiliaceae
  Cây bo

Dalbergia
  Leguminosae
Dipterocarpus
  Dipterocarpaceae
  Dâu
Dipterocarpus dyeri
  Dipterocarpaceae
  Dâu song nang
Dipterocarpus intricatus
  Dipterocarpaceae
  Dâu trai
Dipterocarpus turbinatus
  Dipterocarpaceae
  Dâu

Eucalyptus
  Myrtaceae
Eucalyptus camaldulensis
  Myrtaceae
Excoecaria
  Euphorbiaceae
  Giá
Excoecaria agallocha
  Euphorbiaceae
  Giá

Haemanthus katherinae
  Amaryllidaceae
  African blood lily
Hopea
  Dipterocarpaceae

Hopea odorata
  Dipterocarpaceae
  Sao den
Imperata
  Gramineae
  Tranh
Imperata cylindrica
  Gramineae
  Tranh
Indigofera
  Leguminosae
  Chàm
Irvingia malayana
  Simaroubaceae
  Cay

Jack fruit see *Artocarpus*

Leucaena
  Leguminosae
Lily, African blood see *Haemanthus*

Macaranga
  Euphorbiaceae
Mallotus
  Euphorbiaceae
  Cánh kien
Manihot esculenta
  Euphorbiaceae
  Cassava (manioc), mì
Melaleuca
  Myrtaceae
  Tràm
Melaleuca leucadendron
  Myrtaceae
  Tràm
Mimosa
  Leguminosae

Oxytenanthera
  Gramineae

Pahudia
  Leguminosae
Parinari annamense
  Chrysobalanaceae
  Cây cám
Pennisetum polystachyon
  Gramineae
  Co duôi voi
Phoenix
  Palmae
Phoenix paludosa
  Palmae
  Chà là bien

Pine see *Pinus*
Pinus
  Pinaceae
  Pine
Pterocarpus
  Leguminosae

Randia tomentosa
  Rubiaceae
Rhizophora
  Rhizophoraceae
  Duróc
Rhizophora apiculata
  Rhizophoraceae
  Duróc dôi
Rhizophora mangle
  Rhizophoraceae
  Red mangrove

Saccharum
  Gramineae
Samanea saman
  Leguminosae
  Còng
Sesame see *Sesamum*
Sesamum indicum
  Pedaliaceae
  Sesame
Shorea
  Dipterocarpaceae
  Chai
Shorea thorelii
  Dipterocarpaceae
Sindora
  Leguminosae
  Go mât
Sindora cochinchinensis
  Leguminosae
  Go mât
Sonneratia
  Sonneratiaceae
  Ban

Teak see *Tectona*
Tectona grandis
  Verbenaceae
  Teak, giá ti
Tephrosia
  Leguminosae
Thyrsostachys
  Gramineae
Trema
  Ulmaceae
  Trân mai cong

Mammals

Bear, Asiatic black
  Selenarctos thibetanus
  Ursidae
  Gâu ngua
Bear, Malayan sun
  Helarctos malayanus
  Ursidae
Boar, wild
  Sus scrofa
  Suidae
  Lon rùng
Buffalo, water
  Bubalus bubalis
  Bovidae
  Trâu rùng

Cat, leopard
  Felis bengalensis
  Felidae
  Mèo rùng
Civet, large
  Viverra zibetha
  Viverridae
  Cây dông
Civet, small
  Viverricula indica
  Viverridae
  Cây huong

Deer
  Cervidae
Deer, lesser mouse
  Tragulus javanicus
  Tragulidae
  Cheo

Elephant, Asian
  Elephas maximus
  Elephantidae
  Voi

Gaur
  Bos gaurus
  Bovidae
  Bò tót
Gibbon, white-cheeked (black gibbon)
  Hylobates concolor
  Pongidae
  Vuon

Kouprey
  Bos sauveli
  Bovidae

Langur, douc
  Pygathrix nemaeus
  Cercopithecidae
  Vec ngu sác
Leopard
  Panthera pardus
  Felidae
  Báo
Leopard, clouded
  Neofelis nebulosa
  Felidae
  Mèo gám

Macaque
  Macaca
  Cercopithecidae
  Khi
Mouse, fawn-colored
  Mus cervicolor
  Muridae
Mouse, white
  Mus musculus
  Muridae
Muntjac (barking deer)
  Muntiacus muntjac
  Cervidae
  Mang

Porcupine, brush-tailed
  Atherurus macrourus
  Hystricidae
  Hon
Porcupine, Chinese
  Hystrix hodgsoni
  Hystricidae
  Nhím

Rat
  Rattus
  Muridae
  Chuôt
Rat, Sladen's
  Rattus sladeni
  Muridae
Rhinoceros, Javan
  Rhinoceros sondaicus
  Rhinocerotidae
  Tê giác
Rhinoceros, Sumatran
  Dicerorhinus sumatrensis
  Rhinocerotidae

Sambar
  Cervus unicolor
  Cervidae
  Nai
Serow
  Capricornis sumatraensis
  Bovidae
  Son duong
Squirrel
  Sciuridae
Squirrel, giant flying
  Petaurista petaurista
  Sciuridae
  Sóc bay
Squirrel, striped ground
  Menetes berdmorei
  Sciuridae
  Sóc vàn

Squirrel, striped tree
  Tamiops
  Sciuridae
  Sóc chuôt
Squirrel, tree
  Callosciurus erythraerus
  Sciuridae
  Sóc bung xám

Tiger
  Panthera tigris
  Felidae
  Hô

Zebu
  Bos indicus
  Bovidae

Bird

Pheasant, Edwards's
  Lophura edwardsi
  Phasianidae

Fish

Barilius pulchellus
  Cyprinidae
  Carp

Carp see *Cyprinus*
Chanos chanos
  Chanidae
  Milkfish
Clarius fuscus
  Clariidae
  Catfish
Ctenogobius ocellatus
  Gobiidae
  Goby
Cyprinus carpio
  Cyprinidae
  Carp

Danio regina
  Cyprinidae
  Carp

Glossogobius giurus
  Gobiidae
  Goby

Hampala macrolepidota
  Cyprinidae
  Carp

Mastacembelus armatus favus
  Mastacembelidae
  Spiny eel
Megalops cyprinoides
  Megalopidae
  Tarpon
Milkfish see *Chanos chanos*
Misgurnus fossilis anguillicaudatus
  Cobitidae
  Loach
Mystus nemurus
  Bagridae
  Catfish

Nemacheilus spiloptera
  Cobitidae
  Loach

Ophiocephalus gachua (Channa gachua?)
  Ophiocephalidae (Channidae?)
  Snakehead

190

Ophiocephalus marulius (Channa
 marulius?)
   Ophiocephalidae (Channidae?)
   Snakehead
Oryzias latipes
   Cyprinodontidae
   Killifish
Osteochilus hasselti
   Cyprinidae
   Carp

Puntius partipentazone
   Cyprinidae
   Carp
Puntius semifasciolatus
   Cyprinidae
   Carp

Rasbora lateristriata
   Cyprinidae
   Carp
Rasbora trilineata
   Cyprinidae
   Carp

Scaphiodonichthys
   Cyprinidae
   Carp

Tilapia mossambica
   Cichlidae
   Tilapia

Insects

Fly, fruit
   Drosophila melanogaster
   Drosophilidae

Mosquito, anopheles
   Anopheles
   Culicidae

Thrips,
   Thrips
   Thripidae

Micro-organisms

Achromobacter
   Achromobacteriaceae
   (rod-shaped gram-negative bacteria)

Arthrobacter
   Corynebacteriaceae
   (rod-shaped gram-positive bacteria)

Pseudomonas
   Pseudomonadaceae
   (rod-shaped bacteria)

Streptomyces
   Streptomycetaceae
   (bacteria)

Appendix 3. International Symposium on Herbicides and Defoliants in War: the Long-term Effects on Man and Nature, Ho Chi Minh City, 13–20 January 1983

I. Genesis and sponsorship

This Symposium was conceived by Professor Dr Ton That Tung and Professor A. H. Westing during Tung's visit to Hampshire College (Amherst, Massachusetts, USA) in 1979, and they proceeded to organize the event on an *ad hoc* basis. Following his death in May 1982, Tung was replaced as co-organizer by Professor Dr Hoàng Dình Câu.

Financial support for the Symposium came from the Christopher Reynolds Foundation (New York), Samuel Rubin Foundation (New York), Foundation for Scientific Cooperation with Viet Nam (Fullerton, California), Hampshire College (Amherst, Massachusetts), and several private donors.[1]

Official documents of the Symposium can be found in appendix 4 and in papers 2.A, 3.A, 4.A, 5.A, 6.A, 7.A, 8.A and 9.A.

The Symposium has been described in the literature on a number of occasions (e.g., Carlson, 1983; Hay, 1983; Norman, 1983; Trapp & Kläss, 1984; Vidal, 1983).

II. Governance

Organizing Committee

Professor Dr Ton That Tung (until May 1982)
Professor Dr Hoàng Dình Câu (after May 1982)
Professor Arthur H. Westing

Presidium

Professor Dr Hoàng Dình Câu (Chairman)
Academician Alexander V. Fokin
Academician Vladimír Landa
Professor Paul W. Richards
Professor Arthur H. Westing

[1] Neither the Stockholm International Peace Research Institute (SIPRI) nor the United Nations Environment Programme (UNEP) played a role in either the organization or the financial support of the Symposium. However, the preparation of this volume took place at SIPRI within the framework of the SIPRI/UNEP project on 'Military Activities and the Human Environment'.

Secretariat

Professor Dr Hoàng Dình Câu (Chairman)
Professor Doan Xuân Muou (Scientific Secretary)
Dr Trinh Van Khiêm (Administrator)
Ms Nguyên Ky Minh Phuong (Assistant Administrator)
Professor Arthur H. Westing (Co-ordinator)
Ms Carol E. Westing (Assistant Co-ordinator)

Working Groups

Terrestrial plant ecology and forestry

Professor Dr Thái Van Trùng (Chairman)
Engineer Hoàng Hoè (Vice-chairman)
Professor Arthur W. Galston (Rapporteur)

Terrestrial animal ecology

Professor Vo Qúy (Chairman)
Candidate Dang Huy Huỳnh (Vice-chairman)
Professor Mark Leighton (Rapporteur)

Soil ecology

Candidate Hoàng Van Huây (Chairman)
Candidate Lê Trong Cúc (Vice-chairman)
Professor Paul J. Zinke (Rapporteur)

Coastal, aquatic, and marine ecology

Professor Mai Dình Yên (Chairman)
Dr Bui Thi Lang (Vice-chairwoman)
Professor Samuel C. Snedaker (Rapporteur)

Cancer and clinical epidemiology

Dr Luong Tan Truong (Chairman)
Dr Dô Dúc Vân (Vice-chairman)
Professor Samuel S. Epstein (Rapporteur)

Reproductive epidemiology

Professor Dr Nguyên Cân (Chairman)
Dr Nguyên Thi Ngoc Phuong (Vice-chairwoman)
Dr John D. Constable (Rapporteur)

Experimental toxicology and cytogenetics[2]

Candidate Cung Bính Trung (Chairman)
Dr Alastair W. M. Hay (Rapporteur)

Dioxin chemistry[2]

Candidate Tran Xuan Thu (Chairman)
Professor Christoffer Rappe (Rapporteur)

[2] During the course of the Symposium the Working Groups on 'Experimental toxicology and cytogenetics' and 'Dioxin chemistry' were combined in a formal sense, but functioned *de facto* largely as two groups.

III. Participants

1. Professor Trinh Kim ANH (medicine)
   Director, Cho Ray Hospital, Ho Chi Minh City, Viet Nam
2. Professor Nikolai S. ANTONOV (oncology)
   Ministry of Health, Moscow, USSR
3. Professor Peter S. ASHTON (forest botany)
   Director, Arnold Arboretum, Harvard University, Cambridge, USA
4. Professor Evgeni I. ASTACHKIN (biochemistry)
   Director, Research Institute of Biological Testing of Chemicals, Moscow Region, USSR
5. Dr Doàn Thúy BA (medicine)
   Vice-director, Cho Ray Hospital, Ho Chi Minh City, Viet Nam
6. Dr Tôn That BACH (surgery)
   Viet Duc Huu Nghi Hospital, Hanoi, Viet Nam
7. Dr Bayko D. BAIKOV (ecology)
   Centre of Biology, Academy of Science, Sofia, Bulgaria
8. Dr Luigi BISANTI (epidemiology)
   Istituto Superiore di Sanita, Rome, Italy
9. Dr Valentin A. BOLSHAKOV (soil science)
   Institute of Cell Science, Academy of Agriculture, Moscow, USSR
10. Dr Guozgui BORISSOV (organic chemistry)
    Vice-director, Centre of Chemistry, Academy of Science, Sofia, Bulgaria
11. Dr Eberhard F. BRÜNIG (forestry)
    Director, Institute for World Forestry, Hamburg, FR Germany
12. Professor Lê Van CAN (soil science)
    National Centre for Scientific Research, Hanoi, Viet Nam
13. Professor Dr Nguyên CÂN (surgery)
    Director, Institute for the Protection of Mother and Newborn, Hanoi, Viet Nam
14. Professor Elof A. CARLSON (genetics)
    Department of Biochemistry, State University of New York, Stony Brook, USA
15. Professor Henri CARPENTIER (oncology)
    13, Rue Payenne, Paris, France
16. Professor Dr Hoàng Dình CÂU (surgery)
    Vice-minister, Ministry of Health, Hanoi, Viet Nam
17. Engineer CHAN Tong Yves (botany)
    Ministry of Agriculture, Phnom Penh, Kampuchea
18. Dr Alexei CHESNOKOV (biochemistry)
    Department of Science Organization, Academy of Sciences, Moscow, USSR
19. Dr Nguyên Trân CHIÊN (genetics)
    College of Medicine, Hanoi, Viet Nam
20. Engineer Vo Trí CHUNG (forestry)
    Institute of Forest Inventory and Planning, Hanoi, Viet Nam
21. Dr John D. CONSTABLE (surgery)
    Department of Surgery, Massachusetts General Hospital, Boston, USA
22. Candidate Lê Trong CÚC (soil science)
    Department of Biology, University of Hanoi, Hanoi, Viet Nam
23. Professor Vu Ta CUC (chemistry)
    College of Medicine, Hanoi, Viet Nam
24. Professor Duòng Hông DÂT (agriculture)
    Vice-minister, Ministry of Agriculture, Hanoi, Viet Nam
25. Mr Rendo DAWA (chemistry)
    Institute of Chemistry, Ulan Bator, Mongolia
26. Dr Nguyên Dinh DICH (medicine)
    Cho Ray Hospital, Ho Chi Minh City, Viet Nam
27. Professor Dr Zdenek DIENSTBIER (oncology)
    Faculty of Medicine, Charles University, Prague, Czechoslovakia
28. Engineer Vu Van DUNG (forestry)
    Institute of Forest Inventory and Planning, Hanoi, Viet Nam
29. Dr Daphne F. DUNN (marine invertebrate zoology)
    California Academy of Sciences, San Francisco, USA

30. Professor James H DWYER (statistics)
    Department of Psychology, State University of New York, Stony Brook, USA
31. Professor Samuel S. EPSTEIN (environmental medicine)
    Department of Environmental Medicine, University of Illinois School of Public Health,
    Chicago, USA
32. Dr Karl-Rainer FABIG (medicine)
    Immenhöven 19, Hamburg, FR Germany
33. Academician Alexander V. FOKIN (chemistry)
    Presidium, Academy of Sciences, Moscow, USSR
34. Dr Zoltán FÜLÖP (teratology)
    Department of Anatomy, Medical School, Budapest, Hungary
35. Professor Arthur W. GALSTON (plant physiology)
    Department of Biology, Yale University, New Haven, USA
36. Dr Ruben R. GAVALDA (immunology)
    William Soler Hospital, Havana, Cuba
37. Professor Arne van der GEN (organic chemistry)
    Department of Chemistry, University of Leiden, Leiden, Netherlands
38. Professor Trân Dình GIÁN (ecology)
    Institute of Social Sciences, Hanoi, Viet Nam
39. Professor Haim B. GUNNER (soil microbiology)
    Department of Environmental Sciences, University of Massachusetts, Amherst, USA
40. Candidate Nguyên Van HANH (agriculture)
    College of Agriculture No. 4, Ho Chi Minh City, Viet Nam
41. Professor Maureen C. HATCH (epidemiology)
    Faculty of Medicine, Columbia University, New York, USA
42. Dr Alastair W. M. HAY (biochemistry)
    Department of Chemical Pathology, University of Leeds, Leeds, England
43. Candidate Dinh HIÊP (forestry)
    Institute of Forest Inventory and Planning, Hanoi, Viet Nam
44. Professor Dr Pham Hoàng HÔ (botany)
    Department of Biology, University of Ho Chi Minh City, Ho Chi Minh City, Viet Nam
45. Engineer Hoàng HOÈ (forestry)
    Director, Institute of Forest Inventory and Planning, Hanoi, Viet Nam
46. Professor Phan Nguyên HÔNG (ecology)
    Faculty of Science, Teachers Training College, Hanoi, Viet Nam
47. Professor Vratislav HRDINA (pharmacology)
    Faculty of Medicine, Charles University, Prague, Czechoslovakia
48. Candidate Hoàng Van HUÂY (soil science)
    Department of Pedology, University of Hanoi, Hanoi, Viet Nam
49. Dr Bui Sy HUNG (medicine)
    Director, Tu Du Nhi Dong Hospital, Ho Chi Minh City, Viet Nam
50. Dr Lê Diêm HUONG (epidemiology)
    Tu Du Nhi Dong Hospital, Ho Chi Minh City, Viet Nam
51. Candidate Dang Huy HUỲNH (mammalogy)
    Director, Institute of Biology, National Centre for Scientific Research, Hanoi, Viet Nam
52. Professor Christoph R. JERUSALEM (cytology)
    Laboratory of Cytology, University of Nijmegen, Nijmegen, Netherlands
53. Dr Carl F. JORDAN (ecology)
    Institute of Ecology, University of Georgia, Athens, USA
54. Professor Nguyên Dình KHOA (human ecology)
    Department of Anthropology, University of Hanoi, Hanoi, Viet Nam
55. Professor Mitsushiro KIDA (teratology)
    School of Medicine, Teikyo University, Tokyo, Japan
56. Dr Michail F. KISSELJOV (biochemistry)
    Research Institute of Biological Testing of Chemicals, Moscow Region, USSR
57. Professor Alexi F. KOLOMIETZ (chemistry)
    Institute of Elemental Organic Chemistry, Academy of Sciences, Moscow, USSR
58. Dr Jirí KUCERA (teratology)
    Institute for Mother and Child, Prague, Czechoslovakia
59. Academician Vladimír LANDA (entomology)
    Director, Institute of Entomology, Academy of Sciences, Prague, Czechoslovakia

60. Dr Bui Thi LANG (marine biology)
    Committee for Science and Technique, Ho Chi Minh City, Viet Nam
61. Professor Tôn Dúc LANG (surgery)
    Viet Duc Huu Nghi Hospital, Hanoi, Viet Nam
62. Professor Mark LEIGHTON (animal ecology)
    Department of Anthropology, Harvard University, Cambridge, USA
63. Dr Pham Duy LINH (epidemiology)
    Vice-director, Department of Health, Ho Chi Minh City, Viet Nam
64. Dr Oleg M. LISSOV (chemistry)
    Legal Office, Ministry of Defence, Moscow, USSR
65. Dr Nguyên Xuan LOC (statistics)
    Institute of Mathematics, National Centre for Scientific Research, Hanoi, Viet Nam
66. Dr Zbigniew MAKLES (analytical chemistry)
    Department of Analytical Chemistry, Institute of Hygiene and Epidemiology, Warsaw, Poland
67. Professor Ivan I. MARADUDIN (forestry)
    State Committee of Forestry, Moscow, USSR
68. Dr Lev W. MEDVEDEV (entomology)
    Institute of Evolution, Morphology, and Ecology of Animals, Moscow, USSR
69. Candidate Nguyên Duy MINH (aquatic ecology)
    Faculty of Science, Teachers Training College, Hanoi, Viet Nam
70. Dr Boguslaw A. MOLSKI (forestry)
    Director, Botanical Garden, Academy of Sciences, Warsaw, Poland
71. Professor Isao MOTOTANI (plant ecology)
    University of Agriculture and Technology, Tokyo, Japan
72. Professor Susil K. MUKHERJEE (soil science)
    332 Jodhpur Park, Calcutta, India
73. Dr MY Samedy (radiology)
    Faculty of Medicine and Pharmacy, Phnom Penh, Kampuchea
74. Dr Vjascheslav V. NAZAROV (soil science)
    Ministry of Fertilizers and Pesticides, Moscow, USSR
75. Candidate Bui Van NGAC (botany)
    Ministry of Agriculture, Hanoi, Viet Nam
76. Professor Phung Trung NGAN (ecology)
    Department of Biology, University of Ho Chi Minh City, Ho Chi Minh City, Viet Nam
77. Dr Hô Dang NGUYÊN (epidemiology)
    Director, Provincial Polyclinic Hospital, Tay Ninh City, Viet Nam
78. Dr Kees OLIE (chemistry)
    Laboratory of Environmental and Toxicological Chemistry, University of Amsterdam, Amsterdam, Netherlands
79. Dr OM Sokha (medicine)
    Revolution Hospital, Phnom Penh, Kampuchea
80. Professor Egbert W. PFEIFFER (zoology)
    Department of Zoology, University of Montana, Missoula, USA
81. Dr Pham Hoàng PHIÊT (immunology)
    Cho Ray Hospital, Ho Chi Minh City, Viet Nam
82. Professor Nguyên Hung PHUC (pharmacy)
    Department of Pharmacy, College of Medicine, Hanoi, Viet Nam
83. Dr Nguyên Thi Ngoc PHUONG (surgery)
    Vice-director, Tu Du Nhi Dong Hospital, Ho Chi Minh City, Viet Nam
84. Dr Jaromir POSPISIL (ecology)
    Institute of Landscape Ecology, Academy of Sciences, Pruhonice, Czechoslovakia
85. Dr Yuri G. PUZACHENKO (forest ecology)
    Institute of Evolution, Morphology, and Ecology of Animals, Moscow, USSR
86. Professor Nguyên Huu QUANG (forestry)
    Ministry of Forestry, Hanoi, Viet Nam
87. Professor Vo QÚY (ornithology)
    Department of Biology, University of Hanoi, Hanoi, Viet Nam
88. Professor Vannareth RAJPHO (anatomy)
    Vice-minister, Ministry of Public Health, Vientiane, Laos

89. Professor T. Navaneeth RAO (chemistry)
    Department of Chemistry, Osmania University, Hyderabad, India
90. Professor Christoffer RAPPE (organic chemistry)
    Department of Organic Chemistry, University of Umeå, Umeå, Sweden
91. Professor Paul W. RICHARDS (botany)
    14 Wootton Way, Cambridge, England
92. Professor Dr Slawomir RUMP (toxicology)
    Department of Environmental Toxicology, Institute of Hygiene and Epidemiology, Warsaw, Poland
93. Dr SAU Sok Khonn (medicine)
    Director, 7th of January Hospital, Phnom Penh, Kampuchea
94. Professor Natalio S. SCHARAGER (microbiology)
    Instituto Superior de Medicina, Faculty of Medicine, Havana, Cuba
95. Mr SENG Lim Neou (pharmacy)
    Faculty of Medicine and Pharmacy, Phnom Penh, Kampuchea
96. Professor Samuel C. SNEDAKER (coastal ecology)
    School of Marine Science, University of Miami, Miami, USA
97. Academician Vladimir E. SOKOLOV (mammalogy)
    Director, Institute of Evolution, Morphology and Ecology of Animals, Moscow, USSR
98. Dr Svetlana SOKOLOVA (plant biochemistry)
    Main Botanical Garden, Academy of Sciences, Moscow, USSR
99. Professor Theodor D. STERLING (biological statistics)
    Department of Computing Science, Simon Fraser University, Burnaby, Canada
100. Professor Dang Nhu TAI (organic chemistry)
    Department of Chemistry, University of Hanoi, Hanoi, Viet Nam
101. Academician Armen L. TAKHTAJAN (botany)
    Director, Komarov Botanical Institute, Leningrad, USSR
102. Professor Dr Pham Bieu TAM (surgery)
    Binh Dan Hospital, Ho Chi Minh City, Viet Nam
103. Dr Chanpheng THAMMAVONG (surgery)
    Mahosot Hospital, Vientiane, Laos
104. Professor Ho Si THOANG (chemistry)
    Director, Institute of Chemistry, National Centre for Scientific Research, Ho Chi Minh City, Viet Nam
105. Professor Lê Van THÓI (chemistry)
    Department of Chemistry, University of Ho Chi Minh City, Ho Chi Minh City, Viet Nam
106. Professor Dr habil. Harald THOMASIUS (forestry)
    Forest Management Section, Dresden Technical University, Tharandt, German Democratic Republic
107. Professor Tran The THONG (zoology)
    National Centre for Scientific Research, Hanoi, Viet Nam
108. Candidate Tran Xuan THU (organic chemistry)
    Department of Chemistry, University of Hanoi, Hanoi, Viet Nam
109. Dr Cornelis J. M. van TIGGELEN (clinical medicine)
    Department of Geriatrics, Voorburg Psychiatric Hospital, Vught, Netherlands
110. Dr Károly TÓTH (oncology)
    Research Institute of Oncopathology, Budapest, Hungary
111. Dr Ralf TRAPP (toxicology)
    Research Institute for Chemical Toxicology, Academy of Sciences, Leipzig, German Democratic Republic
112. Dr Dô Thuc TRINH (epidemiology)
    College of Medicine, Hanoi, Viet Nam
113. Candidate Cung Bính TRUNG (genetics)
    College of Medicine, Hanoi, Viet Nam
114. Professor Le The TRUNG (surgery)
    College of Medicine, Hanoi, Viet Nam
115. Professor Dr Thái Van TRÙNG (forest botany)
    Director, Botanical Museum and National Herbarium, National Centre for Scientific Research, Ho Chi Minh City, Viet Nam
116. Dr Luong Tan TRUONG (oncology)
    Director, Institute of Cancer, Ministry of Health, Ho Chi Minh City, Viet Nam

117. Dr Nguyên Anh TUONG (medicine)
Cho Ray Hospital, Ho Chi Minh City, Viet Nam
118. Di Dach Quoc TUYÊN (haematology)
Director, Department of Haematology, Bach Mai Hospital, Hanoi, Viet Nam
119. Candidate Nguyên Van TUYÊN (algology)
Department of Botany, University of Hanoi, Hanoi, Viet Nam
120. Candidate Pham Van TY (microbiology)
Department of Microbiology, University of Hanoi, Hanoi, Viet Nam
121. Dr Dô Dúc VÂN (surgery)
Viet Duc Huu Nghi Hospital, Hanoi, Viet Nam
122. Dr Nguyên Van VAN (surgery)
Viet Duc Huu Nghi Hospital, Hanoi, Viet Nam
123. Professor John H. VANDERMEER (ecology)
Department of Zoology, University of Michigan, Ann Arbor, USA
124. Dr Jules E. VIDAL (botany)
Laboratory of Phanerogamy, National Museum of Natural History, Paris, France
125. Professor Arthur H. WESTING (forest ecology)
Stockholm International Peace Research Institute (SIPRI), Solna, Sweden (also: School of Natural Science, Hampshire College, Amherst, Massachusetts, USA)
126. Dr Nguyên Thi XIEM (medicine)
Vice-director, Institute for the Protection of Mother and Newborn, Hanoi, Viet Nam
127. Professor Mai Dinh YÊN (ichthyology)
Department of Biology, University of Hanoi, Viet Nam
128. Professor Paul J. ZINKE (soil science)
Department of Forestry, University of California, Berkeley, USA

IV. Observers

1. Engineer Phung Tu BÔI (forestry)
Institute of Forest Inventory and Planning, Hanoi, Viet Nam
2. Mr Mohamed S. BOULECANE (agriculture)
Representative, Food and Agriculture Organization of the United Nations (FAO), Hanoi, Viet Nam
3. Candidate Dang Sang CANH (biochemistry)
National Centre for Scientific Research, Hanoi, Viet Nam
4. Mr Nguyên Xuan CU (soil science)
Department of Pedology, University of Hanoi, Hanoi, Viet Nam
5. Dr Do Binh DUONG (gynaeco-obstetrics)
Institute for the Protection of Mother and Newborn, Hanoi, Viet Nam
6. Engineer Vo Trung HANG (forestry)
Provincial Forestry Service, Minh Hai Province, Viet Nam
7. Dr John R. E. HARGER (ecology)
Regional Officer for Science and Technology for Southeast Asia, United Nations Educational, Scientific, and Cultural Organization (Unesco), Jakarta, Indonesia
8. Ms Vi Nguyêt HÔ (Mrs Ton That TUNG) (anaesthesiology)
Viet Duc Huu Nghi Hospital, Hanoi, Viet Nam
9. Candidate Nguyên Dúc KHÁNG (forestry)
Vice-director, Institute of Forest Inventory and Planning, Hanoi, Viet Nam
10. Dr Trinh Van KHIÊM (medicine)
Ministry of Health, Hanoi, Viet Nam
11. Professor Judith L. LADINSKY (public health)
Department of Preventive Medicine, University of Wisconsin, Madison, USA
12. Dr Reynaldo M. LESACA (environmental health)
Director, Regional Office for Asia and the Pacific, United Nations Environment Programme (UNEP), Bangkok, Thailand
13. Dr John H. LEVAN (Le Van HOA) (radiology)
Secretary, US Committee for Scientific Cooperation with Viet Nam, 7558 N. Crawford Ave., Skokie, Illinois, USA

14. Dr John M. LEVINSON (medicine)
    President, Aid for International Medicine, 1828 Wawaset St., Wilmington, USA
15. Dr Nguyên LIEN (epidemiology)
    College of Medicine, Hanoi, Viet Nam
16. Professor Doan Xuân MUOU (virology)
    Ministry of Health, Hanoi, Viet Nam
17. Ms Nguyên Ky Minh PHUONG (pharmacy)
    Ministry of Health, Hanoi, Viet Nam
18. Mr Nguyên Xuân QUÝNH (zoology)
    Department of Biology, University of Hanoi, Hanoi, Viet Nam
19. Candidate Cao Van SUNG (ecology)
    Institute of Biology, National Centre for Scientific Research, Hanoi, Viet Nam
20. Engineer Trân Thi THÁI (biology)
    Department of Haematology, Bach Mai Hospital, Hanoi, Viet Nam
21. Professor Trinh Dinh THANH (forestry)
    Director, Department of Science and Technology, Ministry of Forestry, Hanoi, Viet Nam
22. Professor Trinh Van THINH (veterinary medicine)
    Director, Department of Science and Technology, Ministry of Agriculture, Hanoi, Viet Nam
23. Dr Nguyên Kim TÒNG (gynaeco-obstetrics)
    Institute for the Protection of Mother and Newborn, Hanoi, Viet Nam
24. Candidate Nguyên Hoàng TRÍ (ecology)
    Faculty of Science, Teachers Training College, Hanoi, Viet Nam
25. Dr Nguyên Van TRINH (epidemiology)
    College of Medicine, Hanoi, Viet Nam
26. Candidate Vu Hoài TUÂN (chemistry)
    Department of Chemistry, University of Hanoi, Hanoi, Viet Nam
27. Engineer Nguyên Dinh VINH (forestry)
    Vice-director, Provincial Forestry Service, Dong Nai Province, Viet Nam
28. Ms Carol E. WESTING (special education)
    International School of Stockholm, Stockholm, Sweden

References

Carlson, E. A. 1983. International symposium on herbicides in the Vietnam war: an appraisal. *BioScience*, Washington, **33**: 507–512.
Hay, A. 1983. Defoliants in Vietnam: the long-term effects. *Nature*, London, **302**: 208–209.
Norman, C. 1983. Vietnam's herbicide legacy. *Science*, Washington, **219**: 1196–1197.
Trapp, R. and Kläss, V. 1984. [Delayed harm of the chemical war in Viet Nam.] (In German). *Wissenschaft und Fortschritt*, East Berlin, **34**: 67–69.
Vidal, J. E. 1983. [Ecological consequences of the chemical defoliation in Viet Nam.] (In French). *Bulletin de la Société Botanique de France*, Paris, **130**: 363–369.

Appendix 4. Long-term ecological and human consequences: overall Symposium summary[1]

Hoàng Dình Câu et al.[2]
Viet Nam Ministry of Health

The International Symposium on Herbicides and Defoliants in War: the Long-term Effects on Man and Nature was held in Ho Chi Minh City from 13 to 20 January 1983. Attending the Symposium were 128 scientists from 21 countries (56 from Viet Nam, 72 from elsewhere) and 28 scientific or technical observers, including observers from the United Nations Environment Programme (UNEP), the United Nations Educational, Scientific and Cultural Organization (Unesco), and the Food and Agriculture Organization of the United Nations (FAO). The Symposium was devoted to the long-term effects of the herbicides employed by the US armed forces, with the agreement of the Saigon administration, during the Second Indochina War of 1961–1975.

At the plenary sessions and in the Working Groups the scientists presented and discussed 65 scientific papers (29 Vietnamese, 36 international) dealing with the following four general topics or problems:

1. The scope and nature of the anti-plant chemical warfare programme conducted by the USA primarily in South Viet Nam from 1961 to 1971, so-called Operation Ranch Hand.

2. The long-term effects of the military herbicides on humans (34 reports) and on nature (31 reports).

3. The results of experimental studies on herbicides in the laboratory or, on a small scale, in the field.

4. The results of studies on the consequences of herbicides or related substances from accidents occurring in factories producing them, or else their effects on groups of workers dealing with these chemicals in agriculture or elsewhere.

The scientists present exchanged views and evaluated the results of laboratory studies or of field experiments. They discussed the research work to be conducted in the near future aimed at eliminating the consequences of the indiscriminate

---

[1] Overall summary report of the Presidium of the International Symposium on Herbicides and Defoliants in War: the Long-term Effects on Man and Nature, Ho Chi Minh City, 13–20 January 1983 (appendix 3).
[2] The Presidium consisted of: Hoàng Dình Câu (Chairman), A. V. Fokin, V. Landa, P. W. Richards, and A. H. Westing.

use of herbicides on a large scale. They also discussed the possibilities of inter-national co-operation in the field of scientific research.

During the time the Symposium was held the scientists visited an exhibition displaying the various kinds of chemical weapons used during the war and the effects of the herbicides on humans and nature.

Participants of the Symposium also visited the inland (upland) forest in the Ma Da area of north-western Dong Nai province (in the former Phuoc Long province, Military Region III, War Zone D). Here wartime destruction inflicted upon nature remains very apparent. The Ma Da forest can in effect be considered as one model for experimental field studies on the direct and indirect effects of herbicides on tropical inland forests, the latter including fire. The visit to the Ma Da forest gave participants a clear idea of the lengthy duration of effects of herbicidal disturbance on the natural restoration of tropical inland forests.

During the Symposium the scientists were engaged in active work in a friendly atmosphere. Although most of the scientists met one another for the first time, their discussions and exchanges of views were conducted in an open, straight-forward and frank way; and they worked in their private capacities. The working languages of the Symposium were Vietnamese, English, French and Russian. These factors together helped to ensure good results for the Symposium.

The majority of the participants reached agreement on the following 10 subject areas:

1. Operation Ranch Hand was chemical warfare conducted with herbicides on a large scale in space and time, the first such massive employment in the history of war. It differed completely from failures or explosion accidents in chemical factories. It also differed from the much smaller scale field experiments in other countries or from laboratory experiments. The results of these other occurrences and studies are only of partial usefulness in evaluating what happened to tropical South Viet Nam and to the Vietnamese people during Operation Ranch Hand.

The herbicides employed in Operation Ranch Hand included primarily the following four substances: (a) *2,4-D* (2,4-dichlorophenoxyacetic acid), a total of 26 million kilograms; (b) *2,4,5-T* (2,4,5-trichlorophenoxyacetic acid), a total of 24 million kilograms; containing dioxin (TCDD; 2,3,7,8-tetrachlorodibenzo-*p*-dioxin) as a contaminant in largely unknown amounts, although conservatively estimated to total no less than 170 kg; (c) *Picloram* (4-amino-3,5,6-trichloro-picolinic acid), a total of 1 million kilograms; and (d) *Cacodylic acid* (dimethyl arsinic acid), a total of 3 million kilograms (of which almost 2 million kilograms was elemental arsenic). It must be noted that the presented amounts of chemicals expended, and the amount of the dioxin contaminant, are derived from official US sources for which there is no independent source of verification.

These four chemicals were applied primarily in the following three mixtures: (a) *Agent Orange*, a mixture of 2,4-D and 2,4,5-T (and thus containing a trace of dioxin); (b) *Agent White*, a mixture of 2,4-D and picloram, and (c) *Agent Blue*, cacodylic acid.

2. Over the last two decades, many experimental studies on herbicides have been conducted in the research facilities of many countries. No full agreement has yet been reached on the results and conclusions regarding the effects of these chemicals on experimental animals. However, through many years of research with admirable patience and increasingly precise methods, the majority of scientists recognize that phenoxy and certain other herbicides used at a high dose or at a low dose for a long period of time will affect animals adversely. They may be variously mutagenic (cause genetic, i.e., DNA or gene, damage), carcinogenic (produce cancers), or teratogenic (result in birth defects).

3. Studies on workers in factories producing herbicides have also been conducted over the last few years. Those studies confirm the toxicity of herbicides, especially of 2,4,5-T and dioxin. The signs of immediate and long-term poisoning due to chlorophenoxy-acetic substances have been described in the medical literature. The manifestations considered as characterizing such poisoning are: chloracne (acne-like skin eruption due to contact with certain chlorinated compounds); porphyria cutanea tarda (symptomatic disorder of porphyrin pigment metabolism); asthenia (weakness), and so on. The reactions to these pathogenic agents differ from one individual to another, as do the manifestations of these reactions, which renders statistical analysis or other evaluation difficult.

4. The Symposium reserved most of its time for the evaluation of the long-term effects of chemical warfare with herbicides in southern Viet Nam. Foreign scientists attending the Symposium highly valued the contributions made by Vietnamese scientists who, despite the limited facilities and other difficulties during and after the war, were largely able to overcome these problems and make valuable research contributions. Indeed, the reports and suggestions made by Vietnamese scientists at the Symposium provided the crucial basis for discussion at the plenary sessions and in the Working Groups. The large-scale field studies carried out by Vietnamese scientists in various localities in both southern and northern Viet Nam have provided much information of scientific value not previously available from other countries.

5. Nature and natural resources in Viet Nam have been substantially damaged. This destruction is for a complexity of reasons. However, the participants agreed that the most important cause of this extensive damage to nature was the large-scale use of herbicides. Immediately after the spraying, the toxic substances exerted their direct destructive effects on the vegetation and, to some extent, on the animals living in the inland, coastal, aquatic and marine habitats. The direct and indirect repercussions of these immediate effects have lasted to this day. Time has only slowly helped to reduce the severity of the deleterious effects. Restoration can only be a slow process, occurring most readily on very small areas. Photographs taken from the air or space have only partially reflected the true state of the damage that still remains in the sprayed tropical forests.

6. The herbicides, sprayed on a large scale, at a high concentration, and in large amount, have changed the composition of some soils, destroyed useful micro-organisms, and, in some instances, caused the soil to lose fertility and to deteriorate in other ways. Many areas which had been covered with trees and other woody plants have become savannas of low productivity. They contain

only wild grasses or a number of secondary successional plants having little economic value, and support rodents which are disease carriers. Evidence from aerial photography and other sources indicates that some of these savannas are continuing to expand in size. Some species of valuable tropical wood are facing the danger of extermination, as are some other important terrestrial and aquatic plants and animals. Transforming these savannas back to forest or building them into new economic zones for agriculture presents difficult problems, the solutions of which are often far beyond the present abilities of the Vietnamese people. The various impacts on nature have in many instances undermined the whole rural human life-support system.

7. The herbicides sprayed on the land were washed away to lowland areas, often far from the sprayed areas, and decomposed over varying periods of time. The most dangerous among them was Agent Orange, which was widely used from 1962 to 1970. As noted earlier, Agent Orange contains the impurity dioxin, an extraordinarily toxic and very resistant substance that persists for a long time in nature. According to official (unverifiable) US sources, some 72 million litres (81 million kilograms) of the various herbicides were sprayed, of which Agent Orange accounted for 44 million litres (57 million kilograms) containing 24 million kilograms of 2,4,5-T. At this time, the most important thing to determine is whether there still exists dioxin in nature in Viet Nam. To this end, analyses were made in 1981 of 11 soil samples taken in a rural area within the administrative unit of Ho Chi Minh City, at different depth levels. In one sample, taken at a depth of 1 m, there was a trace of dioxin, that is, a concentration of about 1 ng/kg (ppt) of soil. In another, taken at the soil surface, the concentration was found to be 16 ppt.

8. There are not as yet many scientific studies identifying the biological (ecological) cycle of dioxin from the soil into plant species, animal species, food, and people. Dioxin and the decomposition products of the herbicides sprayed have probably been carried to lowland areas in Viet Nam and neighbouring countries and surrounding seas. Where will these substances end up? How will they be decomposed? What dangers will they present? How soon will the dioxin be decomposed to insignificance? These points have not yet been established. The opinions put forth at the Symposium were only estimates which must be improved upon and verified over a long period of time.

9. The evaluation of the long-term effects of herbicides is a most difficult and complex task. It is therefore difficult to reach full agreement, since the conditions under which scientists work differ from one country to another. However, most of the conclusions drawn by Vietnamese scientists have corroborated the results of experiments conducted by many scientists elsewhere in the world. Reports by Vietnamese scientists have suggested that the herbicides affected chromosomes and that they caused congenital abnormalities (birth defects), molar pregnancies (in which a hydatidiform mole develops in lieu of a foetus), and choriocarcinomas (cancers of the membrane surrounding the foetus). Vietnamese veterans of the war exposed to the toxic chemicals for a long time during the war years may pass on such abnormalities to their offspring. The rate of monsters born to families of Vietnamese veterans of the war seems to be higher

than in normal families. The Vietnamese studies also provide some data on how these chemicals affect human health and how they cause cancer. Herbicides penetrating human bodies may cause long-term effects even though the victims have already left the contaminated areas.

Many of the preliminary conclusions of the Vietnamese scientists are new points. They were observed in Vietnamese society and have never been dealt with, or else only inadequately so, in foreign research works.

10. During the Symposium scientists also agreed upon the following final three points:

(*a*) Further studies should be continued for many years on the long-term effects of herbicides used in the war on humans and nature in Viet Nam.

(*b*) International co-operation between Vietnamese scientists and their foreign colleagues is necessary to promote study and to determine the effects of herbicides, and to find measures to cope with them, both in the interest of the Vietnamese people and of other peoples elsewhere in the world. Thus this Symposium has had a humanitarian aspect, serving the interests of all people.

(*c*) Measures to cope with the effects of herbicides are complicated and difficult. They involve many fields of science, technology, economics, management, and culture; and they call for appropriate governmental policies. Clearly, they require a high level of science divorced from politics, the co-operation and commitment of the entire population, and substantial investments of money and matériel. Unrestricted assistance from the international community in all fields related to this endeavour is an urgent necessity.

In closing, it should be noted that the following additional documents provide official summaries of the eight Working Groups of the Symposium: (*a*) Terrestrial plant ecology and forestry (paper 2.A); (*b*) Terrestrial animal ecology (paper 3.A); (*c*) Soil ecology (paper 4.A); (*d*) Coastal, aquatic, and marine ecology (paper 5.A); (*e*) Cancer and clinical epidemiology (paper 6.A); (*f*) Reproductive epidemiology (paper 7.A); (*g*) Experimental toxicology and cytogenetics (paper 8.A)[3]; and (*h*) Dioxin chemistry (paper 9.A)[3].

[3] During the course of the Symposium the Working Groups on 'Experimental toxicology and cytogenetics' and 'Dioxin chemistry' were combined in a formal sense, but functioned *de facto* largely as two groups.

Index